T0300107

Paradigms of Titanate Centered Energy Materials

Titanate-centered perovskites are an important class of crystalline materials with outstanding structural and property tunability. Recently, titanate perovskites have been extensively investigated as new-generation materials for various energy-allied applications, together with photovoltaic cells, light emitting diodes, photo detectors and lasers, for their exceptional optical and electronic properties. With increasing attention on the development of nanotechnology and multidisciplinary research, scientists have been trying to downscale the titanate perovskite structures into the nanoregime, so as to further boost their performances and applications. Compared with bulk perovskite materials, perovskite nanomaterials exhibit a series of advantages, such as fabrication of thin films and flexible devices and PNMs display high process ability. Perovskite nanomaterials feature rich and controllable facets and active sites. Besides benefiting from the small-size effect and quantum effect, perovskite nanomaterials are endowed with outstanding photo-electromagnetic properties. Furthermore, the applications of titanate-centered perovskites in the fields of energy, environment, WLEDs, forensic science and piezoelectric devices will be reviewed. This book will also discuss the possible solutions to the problems in their application by optimizing their compositions, structures and preparation methods.

This book will systematically summarize the key points in the design, synthesis, property improvements and application expansion of titanate-centered perovskites. The different perovskite structures and the rational design of functional materials will be discussed in detail. The advantages, disadvantages and experimental parameters of different synthesis methods for titanate-centered perovskites will be reviewed. This book will also summarize some practical experiences in selecting suitable elements and designing multifunctional materials according to the mechanisms and principles of elements promoting the properties of perovskites by covering almost last 20 years of literature. At the end of this book, we will provide an outlook on the opportunities and challenges in the view of development in titanate-centered perovskites.

The inclusive effectiveness and practicality of prospective titanate energy materials and systems are directly connected to many materials-related factors. This volume of book features totally seven sections that envelop a wide range of titanate energy materials. They cover the modern developments involving materials for alternative and renewable energy sources and systems, including energy storage and batteries, nanocomposites, hydrogen, solar, wind, geothermal, biomass and nuclear. The book is a significant orientation for students and researchers (from academics, but also industry) interested in understanding the properties of emerging titanate energy materials and challenges in the recent era.

Paradigms of Titanate Centered Energy Materials

Sanjay J. Dhoble, Abhijeet R Kadam,
G.P. Darshan, S.C. Sharma,
and H. Nagabhushuna

CRC Press
Taylor & Francis Group
Boca Raton London New York

CRC Press is an imprint of the
Taylor & Francis Group, an **informa** business

Designed cover image: © Shutterstock, ID 2164958975, Peter Hermes Furian

First edition published 2023
by CRC Press
2385 NW Executive Center Drive, Suite 320, Boca Raton FL 33431

and by CRC Press
4 Park Square, Milton Park, Abingdon, Oxon, OX14 4RN

CRC Press is an imprint of Taylor & Francis Group, LLC

Library of Congress Cataloging-in-Publication Data
Names: Dhoble, Sanjay J., 1967- author. | Kadam, Abhijeet R., author. |
Darshan, G. P., author. | Sharma, S. C. (Mechanical engineer), author. |
Nagabhushuna, H., author.
Title: Paradigms of titanate centered energy materials / Sanjay. J Dhoble,
Abhijeet R. Kadam, G.P. Darshan, S.C. Sharma, H. Nagabhushuna.
Description: Boca Raton : Taylor and Francis, 2024. |
Includes bibliographical references and index. | Identifiers: LCCN 2023003478 (print) |
LCCN 2023003479 (ebook) |
ISBN 9781032464664 (hardback) | ISBN 9781032464848 (paperback) |
ISBN 9781003381907 (ebook)
Subjects: LCSH: Perovskite materials. | Titanates. |
Energy storage—Materials. | Energy conversion—Materials.
Classification: LCC TA455.P46 D46 2024 (print) | LCC TA455.P46 (ebook) |
DDC 621.042—dc23/eng/20230503
LC record available at https://lccn.loc.gov/2023003478
LC ebook record available at https://lccn.loc.gov/2023003479

ISBN: 9781032464664 (hbk)
ISBN: 9781032464848 (pbk)
ISBN: 9781003381907 (ebk)

DOI: 10.1201/9781003381907

Typeset in Palatino
by codeMantra

Contents

Authors

Prof. Sanjay J. Dhoble obtained his M.Sc. degree in Physics from Rani Durgavati University, Jabalpur, India in 1988. He obtained his Ph.D. degree in 1992 in Solid State Physics from Nagpur University, Nagpur, India. Dr. Dhoble is presently working as a Professor in the Department of Physics, Rashtrasant Tukadoji Maharaj Nagpur University, Nagpur, India. He has about 30 years of experience in teaching at undergraduate and post-graduate levels and 29 years of research experience. Consecutively in the years 2020 and 2021, Dr. Dhoble stood among the "Top 2% Scientists in the World" in the prestigious list published by a team of scientists in Stanford University, USA, in one of the highly rated journals, *PLOS Biology*. He also received the Best Research Award-2016 for the outstanding contribution in the field of research and related extenuation activity for the university, presented by R.T.M. Nagpur University, Nagpur, India on 5 September 2016. He is also a recipient of India's Top Faculty Research Award-2018 by Careers 360 for the top ten researchers in India in Physics. During his research career, he has worked on the synthesis and characterization of solid-state lighting materials and phosphors for solar cell efficiency enhancement, as well as development of radiation dosimetry phosphors and biosynthesis of nanoparticle and their applications. Dr. Dhoble published 44 patents, and few of them are already granted (Granted patents are 18: 12 Australian, 5 Indian and 1 South Korean). More than 717 research papers by him are published in Scopus indexed journals, and his h-index is 37 and has 8131 citations on Scopus. Dr. Dhoble is an Editor of *Luminescence: The Journal of Biological and Chemical Luminescence*, John Wiley & Sons Ltd. He is also a Fellow of the Luminescence Society of India (LSI) and Fellow of Maharashtra Academic of Sciences (MASc).

Dr. Abhijeet R Kadam is the research scholar at the Department of Physics, OP Jindal University, Raigarh. He completed his Ph.D. in Physics under the supervision of Dr. Girish C. Mishra and co-supervision of Dr. S. J. Dhoble. He completed his M.Sc. from Dr. Ambedkar College, Nagpur in 2018. Dr. Abhijeet R. Kadam also worked on a major research project as a junior and senior research fellow entitled "Downshifting compatible nano materials for improving the spectral response of present and future generation solar cells" vide Letter no. [DST/NM/NS/2018/38(G) dated 16 JAN 2019] from June 2019 to March 2022. His area of interest includes photoluminescence, material science, nanomaterial synthesis, thermoluminescence dosimetry, photovoltaic efficiency enhancement and mercury extraction. He is a life member of Indian Science Congress Association (Membership No. L42203). During his research work, he published 9 patents, out of which 3 Australian patents,

1 South African patent and 1 Indian patent received grant. Dr. Kadam published 40 research papers, out of which 39 papers are in Scopus indexed journal. He is reviewer of many reputed publishing houses like Wiley, Elsevier, American Chemical Society, etc. Moreover, 8 of his book chapters are published in reputed international books from NOVA Publications and Woodhead Publishing (Elsevier).

Dr. G.P. Darshan obtained his Ph.D. degree (Physics) from Bharathiar University (Coimbatore, India) in 2019. He is presently working as an Assistant Professor in the Department of Physics, Faculty of Mathematical and Physical Sciences, M.S. Ramaiah University of Applied Sciences, Bangalore, India. He has more than 10 years of teaching and 8 years of research experience. He contributed immensely toward the development of eco-friendly, low-cost and efficient rare earth-doped phosphors which have high luminous efficiency, low power consumption, long operational lifetime, durable and eco-friendly. The chemistry of metal oxides, silicates, molybdates, tungstates and various biocompatible polymer host matrices was altered through rare earth doping for efficient light harvesting. Many of these could be potential commercial materials for industrial WLED's applications.

Dr. S.C. Sharma is presently working as the Director of National Assessment and Accreditation Council. He is an eminent Professor and Researcher in the field of Mechanical Engineering. Under his guidance, 29 sponsored projects were completed and 36 students were awarded with Ph.D. degree. He has authored/coauthored 18 books and 8 book chapters. His research focused on basic engineering aspects and worked on structural and mechanical properties of aluminum, zinc, graphite, cast iron-based alloys and composites, corrosion behavior, tensile strength, fracture toughness and heat treatment. Later, his work turned toward the advanced materials and their composites to meet the industry needs. Nanomaterials and their composites prepared by his research group find the latest applications in various disciplines. He is a Member of Indian Ranking Society (IRS), New Delhi, Indian National Commission for Cooperation with UNESCO (INCCU), New Delhi and a Life Member of Institute for Social & Economic Change, (ISEC) Bangalore. He was in the Search Panel to select Vice Chancellors for central and state-supported universities.

Dr. H. Nagabhushuna obtained his M.Sc. degree in Physics from Bangalore University, Bangalore, India in 1998. He obtained his Ph.D. degree in 2003 on Solid State Physics from Bangalore University, Bangalore. Since 2015, he has worked as Chairman and Professor in the Department of Physics, Tumkur University, Tumkur, India. During his research career, he is involved in the synthesis and characterization of nanophosphors for the development of optoelectronic devices, anti-counterfeiting and finger print applications. Dr. H. Nagabhushana published more than 430 research papers in international journals and more than 200 national journals. His academic achievement includes the best research paper awards for more than 20 research papers in conferences and reviewer of more than 100 international journals. He has a teaching experience of 20 years and has visited various countries. He has successfully guided 30 students for Ph.D. degree and completed 9 research projects. He is a coauthor of 15 books/chapters. He is the editorial board member of various journals. He is a life member of Luminescence Society of India, Electrochemical Society of India and Material Research Society of India.

1

Introduction

1.1 Introduction

1.1.1 Perovskite Materials

Electrical energy provides crucial support for human being's living [1]. The energy demand of human communities is increasing with the increase in population all over the world. A key source of energy at present is fossil fuels, which are gradually dwindling and causing numerous ecological exertions [2]. Solar energy is the leading energy source on earth. Endeavors have been made to modernize solar energy into electrical energy by using solar panels, for instance. Regardless, this energy source is confined by its daytime maneuver, low energy density, and inappropriate storage and transportation [3–5]. Another way to utilize solar energy is to switch the energy into chemical-bond energy in a process imitating natural photosynthesis. In the artificial photosynthetic process [6], direct sunlight is used to split water molecules into hydrogen and oxygen. Hydrogen gas has high-density chemical-bond energy and has been documented as a capable energy carrier. In the course of fuel-cell systems, hydrogen can produce electricity without causing contamination during the electrochemical exchange process [7,8]. Moreover, hydrogen can be stored, making it easy to transport and appropriate as a medium fuel that can substitute presently used fossil fuels [9].

Perovskite is a mineral that was primarily discovered by Gustav Rose in the 1830s and is named after the Russian mineralogist A. Von Perovski [10]. Since the 1940s, a large amount of work has been devoted to the structural, chemical, and physical properties of perovskite in order to derive materials for modern applications. These are the class of oxides with repeated lattice structures of ABX_3 or A_2BX_4, where A and B are cations and X is the anion that ties A and B together via the formation of an ionic bond. The ionic radius of A is larger than that of B. The structure of perovskite materials can be viewed as follows: A is located on the edge, and B is located at the center of an octahedron. The cationic radii are $r_A > 0.09$ nm and $r_B > 0.051$ nm [10,11]. In view of the probable valence of the cations and the electro neutrality of the material, various charge distributions can be experienced in the structure. Hence, more than 90% of the metallic elements in the periodic table can be utilized to constitute perovskite structures. One of the major benefits of perovskite structures is the possibility to espouse a wide range of different compositions, by changing either the A or B cation or being partly replaced by other cation(s) of similar or diverse valence, resulting in a common formula of $A_{1-x}A'xB_{1-y}B'yO_{3\pm\delta}$, which can modify the surface, bulk, and redox properties. Geometrical restrictions of dodecahedral or octahedral cavities are the reason behind the stability of the structure. Moreover, the materials with formulae of AB_2X_4 (A_3X_4 when A=B) such are documented as a spinel structure possesses moderately comparable physicochemical properties and are used extensively simultaneously with ABX_3 for large catalysis applications caused by their high activity and immovability. Anion X is oxygen or non-oxygen materials, for instance fluorine (–F) or methyl (–CH_3). Notwithstanding the fact that most perovskite materials have a bandgap greater than 3 eV, a wide selection of doping segments allows further changes of the energy bandgap, and these customized materials can be used as materials with high photocatalytic efficiency [12,13].

Perovskite oxides are possibly used in different kind of reactions, such as solid and gas reactions at higher temperatures or liquid reactions at room temperature, or then again even those led under exposure environment [14,15]. It can be concluded from the structure that perovskite oxides are materials comprising two or more simple oxides with high melting points. As-arranged perovskites confirm low synergist effectiveness in the related responses since high efficiency necessitates a high surface territory for contact with the substrate. Perovskite oxide materials with various textural structures, particularly permeable models, have huge surface zones and show upgraded execution. Such materials have numerous worthwhile attributes and unit operations that empower them for a wide margin of synergist applications. They display structural tenability which is effectively attainable by means of templating and described by a permeable system where numerous variables which are to be controllable, together with divider thickness, pore interconnectivity, shape and pore size. Porous perovskite oxides empower great supervision in the arrangement and integration of utilitarian materials (e.g., metal nanoparticles), which is significant for accomplishing superior catalytic movement, selectivity, and immovability of the resulting catalysts in addition to fast mass transfer.

In prior work, perovskite materials were prepared by the conventional solid-state reaction method [16]. However, the prepared phosphors were not homogeneous and had impurity in it. The materials were also very sensitive to thermal changes, the amount of starting material and the synthesis method. A number of new methods have been used to conquer defects such as the glycine–nitrate route [17], wet-chemical method [18], freeze drying method [19], and sol–gel method [20]. Recently, perovskite materials have been incorporated into thin films and nanostructures using modern synthesis techniques such as chemical conversion hydrothermal synthesis [21,22], laser-assisted chemical vapor deposition [23,24], and facile thermal treatments [25–27].

1.1.2 Energy Exclamations Today

One of the supreme challenges for humankind in the 21st century is how to save fossil fuels as the primary sources of energy which are diminishing gradually [28]. Fossil fuels are non-renewable forms of energy sources which took several thousands of years to form [29]. Therefore, their reserves are exhausted more rapidly than the rate at which new ones are formed and/or discovered. While the concern over fossil fuel materials is habitually the direct or indirect cause to provincial and universal clash, the fabrication, diffusion, and utilization of fossil fuels also lead to ecological deprivation [30]. Combustion of carbon-based fossil fuels produces not only air pollutants, such as sulfur oxide and heavy metals, but also CO_2, the notorious conservatory gas broadly whispered to be the offender of comprehensive climate change [31]. One of the solutions to this energy challenge, on the one hand, rely upon good organization in fabrication, transmission, and exploitation of the remaining fossil fuels while reducing the negative contact to the atmosphere. One the other hand, technologies and transportation have to be urbanized or improved for the smooth evolution to the alternative and renewable energy sources [32,33] such as nuclear power, solar energy, wind energy, geothermal energy, biomass and bio fuels, and hydropower.

The hasty development of nanoscience and nanotechnology (the study and control of materials and phenomena at length scale of 1 to 100 nm) since the past couple of decades demonstrated that nanotechnology holds the key to much technological advancement in the energy region which confides, leastways some context, on having novel materials with greater qualities. The most promising applications of the field of nanotechnology or nanomaterials for the energy creation field will be in photovoltaics and hydrogen conversion. Nanotechnology arbitrates at several phases of energy flow that starts from the crucial energy source and ends up at the last user. The rising and disperse edge of what can be disputably energy and the multifarious flows of energy in the public and environment make it impractical to draw an ambivalent classification of energy production. Here the phrase 'energy production' comprises all procedures that transfer energy from essential energy sources to optional energy sources. It does not cover energy production measures that are related with non-renewable energy sources, i.e. fossil fuels and a few renewable sources such as ocean wave energy, wind energy, and hydropower.

1.1.3 Titanate Materials in Energy Production

The problem with energy depletion and ecological effluence has now turned out to be a hot topic globally in past few years [34,35]. Grasping utilization of carbon-based fossil fuel energy resources has allow to carbon dioxide accretion in the environment, with the production of global warming and environmental effect [36]. The energy and ecological disturbed appeal to the renovation of our energy formation to depend on environmental and sustainable roots, for example, solar energy [37–39]. Furthermore, air and water, the most essential matter on which human existence relies on, has turned into a hazard for humans owing to the remission of chemical wastes [40,41]. Dirty water is frequently polluted by organic dyes that are barely recyclable. The Fukushima disaster of Japan in March 2011 discharged radio-active ions into nature [42]. In the midst of amplified social impact and common community consciousness, new compounds or approaches are in urgent requirement to resolve or moderate these inconveniences.

Due to its constructive electronic and opto-electrochemical properties, titanium dioxide is perhaps the most vigorously examined oxide material in tending to energy and ecological disaster. Owing to these properties of titanate materials, it has been extensively used in solar cells, photocatalysts, lithium-ion battery electrodes, smart coatings, finger prints, etc. [7,43–46]. Nonetheless, enhanced properties are essential to convene high demands and multifaceted desires in the society. A number of attempts have been made to create nanostructured titania materials with designed syntheses, morphologies, or heterostructures. The successful improvement of titanium dioxide nonmaterial has also prospered the research of a class of TiO_2-based structures: layered titanate materials [47,48].

Layered titanates comprise close structural similarity to titanium dioxide, both formed from TiO_6 octahedral units associated with involvement of bends and edges [49,50]. Undeniably, the dissimilarity among layered titanate and TiO_2 is not exceedingly apparent, in particular, in the previously reported work [51], and the term titanium dioxide is notwithstanding rarely familiar with name the material that should even more conventionally be known layered tita-nate. Exceptionally large ion-exchange capability [52–54], fast ion diffusion and interpolation, high surface charge density [55], and ability to transform [29,56] are some attractive features of layered titanate materials. These advantageous properties of layered titanate are furnished by their inimitable crystal structure, where the negatively charged two-dimensional titanium comprising sheets are detached to a huge distance by negotiable cations and molecules in the interlayer [57–59]. These features of layered titanate materials allow the use of titanate as a supple precursor for TiO_2 nano-engineering and in energy production [60–63]. The ability of construct titanate in different shapes permits the morphology to be innate by the consequent TiO_2 formation. There are frequent reviews on the modern advancement of titanate nanotubes or the resulting titania for its photocatalytic application [64,65]. The open crystal structure of layered titanate authorizes simple and consistent doping of atoms, which is arduous to attain if functioning precisely with the more compacted TiO_2 material [66].

1.1.4 Crystal Structure of Titania and Titanate in Energy Materials

Titania can subsist in several forms of crystalline structure; however, it is mainly explained in three general forms of crystal structure namely rutile, anatase, and brookite as displayed in Figure 1.1.

(a) (b) (c)

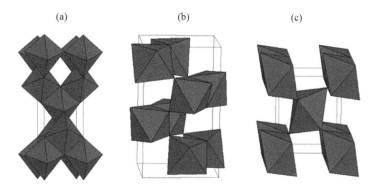

FIGURE 1.1
Crystal structures of TiO$_2$ structure: (a) anatase, (b) brookite, and (c) rutile phase of TiO$_2$. (Reprinted with the permission from Ref. [61] Copyright © 2018 Elsevier publications.)

In all structures, titanium cations are six-fold coordinated to oxygen anions, forming distorted TiO$_6$ octahedra, and all structures have TiO$_6$ octahedra connected by sharing the octahedral edges. Dissimilar crystal structure of titania materials differs by the dimensional structure of TiO$_6$ octahedra construction obstructs. Rutile is the thermodynamically stable form of bulk titania, which has tetragonal structure with a space group P42/mnm (136). Even though being meta-stable at the bulk form, anatase and brookite can be stable when the crystal is in nano-size, by reason of their lesser surface energy [67]. Shape and size of nanoparticles are depending on the conditions such as synthesis technique, precursors, and temperature variation during synthesis. Previous observation of crystal structure during synthesis by the researchers is reported. In the study of Zhang et al., he observed that when the heating temperature was at 450°C, only anatase and brookite were present in the mixed phase particles, and their sizes were diverse from approximately 10 to 14.5 nm and from approximately 12 to 17 nm, respectively, when heating time increased from 2 to 24 h. Rutile phase commenced to form after heating the anatase and brookite mixture at 580°C, and the rutile particle develop at about 60 nm after 21 h [68].

In addition to rutile, anatase, and brookite, there is one interesting phase of pure titania, TiO$_2$(B). This phase is habitually created by means of solution synthesis. Even though having the same chemical formula TiO$_2$, TiO$_2$(B) acquires a very open crystal structure and contributes to further resemblance with titanate apart from TiO$_2$ phases [69]. Its space group C2/m (14) is similar to monoclinic titanate [70] and can be predictable as a titanate sheet encompassing structural unit composed of merely two TiO$_6$ octahedra, with contiguous sheets connected together via edge-sharing. Apart from precipitating directly from solution, TiO$_2$(B) is recognized as an intermediary product in the calcinations of titanate, subsequent to the modified evolution of protonating titanate TiO$_2$(B) and anatase TiO$_2$ [71,72]. Anatase and rutile are the most investigated polymorphs of titania for solar power applications, for instance, photocatalyis and photovoltaic [73,74]. Each of them encompass an indirect band gap, which is 3.0 eV for rutile and 3.2 eV for anatase, equivalent to ultra-violet (UV) light absorption. It is usually acknowledged that anatase is the most energetic phase for photocatalysis applications, owing to its enhanced electronic as well as surface chemistry assets [75,76]. Their properties can be more superior by doping or forming heterojunctions among further phases with constructive electronic coupling.

In recent innovation, these TiO_2-based materials have been errone-ously allocated as anatase TiO_2 although later studies inveterate that they were in fact a category of layered titanate material with a general formula $M_xTi_yO_{x/2+2y} \cdot zH_2O$ (M=H, Li, Na, K, etc.) [77–79]. Titanate materials can be divided into two subclasses on the basis of crystal symmetry, viz monoclinic and orthorhombic. All the monoclinic forms of titanate belong to space group C2/m (12) [80] while orthorhombic titanate are with space group Immm (71), which is normally known as lepidocrocite titanate [81]. The assorted stoichi-ometry of monoclinic titanate is associated to the number of TiO_6 octahedra creating the structural unit. As in $H_2Ti_3O_7$, the structural unit is composed of 3 octahedra blocks [82], while in $H_2Ti_4O_9$, the structure unit has 4 octahedra blocks [83]. The layering environment monoclinic titanate can also be preju-diced by means of dimension and quantity of the further cations compared with titanium. When there is no adequate amount of cations to sustain the large interlayer spacing, the distinct interlayers will stick together by allo-cating to TiO_6 octahedra corners. This can be useful for the management of radioactive ions. A different layered variety of titanate structure is the ortho-rhombic structure. Even though it lives in a diverse crystal system, the TiO_6 octahedra arrangement into orthorhombic titanate is intimately associated to monoclinic titanate. Undeniably, orthorhombic titanate can be conceptual-ized as the monoclinic titanate with the structural unit composed of count-less number of TiO_6 octahedra. Dissimilarity among the monoclinic and orthorhombic titanate is habitually complicated since their crystallinity is typically low, as well as their structural resemblance confided close attribute peak positions in X-ray diffraction (XRD) [84].

1.1.5 Importance of Energy Sources and Energy Materials

One of the grand problems in the 21st century is undeniably energy storage [85]. Due to the requirements of current society and rising natural problems, it has become necessary that novel low-cost and ecologically well-disposed energy conversion and storage systems are established, thus being the fast improvement of research in the area [86]. The functioning of these types of devices relies confidentially on the potential applications of their materials. Inventive material science lies at the center of the approaches that have just been made in energy conversion and storage, for example, the initiation of the rechargeable lithium battery [87,88]. Additional development in materials, not progressive changes, hold the way to new ages of energy storage and conversion devices. Nanostructured materials have pulled in incredible enthusiasm for last few years on account of the surprising mechanical, electrical, and optical properties capable of binding the dimensions of such materials and in view of the blend of bulk and surface properties to the overall performance [89]. One needs to just think about the stunning advancements in microelectronics to welcome the capability of materials with decreased measurements. Nanostructured materials are ending up progressively significant for electrochemical energy storage [90–92]. Here we address this subject. It is critical to welcome the preferences and detriments of nanomaterials for energy conversion also, capacity, just as how to control their combination and properties. This is a generous test confronting those associated with materials investigation into energy conversion and storage.

As one of the basic concerns, energy straightforwardly creates consequences in the financial system, the earth, and the security of people [93–95]. Since old occasions, progresses in the improvement of materials and energy have characterized and restricted human social, mechanical, and political desires. In today's modern world, with instant worldwide correspondence and the rising desires of developing nations, energy challenges are more prominent than at any other time seen previously. Access to energy is the basic to wealth, way of lifestyle, and self-image of each nation. Current utilization of non-renewable energy sources by low-proficiency engines in power age and transportation essentially adds to financial issues and ecological issues including conceivable environmental change [96,97]. Enhancing rates of energy utilization around the globe have prompted a relating ascend in worries about where the sustainable energy originates from and what sort of future energy frameworks ought to be developed. A large number of material-based solutions have been investigated for empowering different energy advancements to accomplish available, inexhaustible, and economical energy avenues for future generation [64]. Novel materials and advancements are required for propelled energy devices and inventive ways to deal with collecting, dispersing, and storing energy.

Energy resources might be delegated capital assets, appropriate for ultimate use devoid of transformation to a further structure, or inferior resources, where the functional type of energy requires considerable change from capital assets [98]. Wind power, fossil fuels, wood fuel, solar power, and fossil fuels such as coal, oil, natural gas, and uranium are the main examples of primary energy sources [2,99–102]. Electricity, hydrogen, or other synthetic fuels are examples of secondary energy sources. Amidst all those energy resources, renewable resources are those that recuperate their ability in a time noteworthy by human needs [103]. Hydroelectric power or wind control, at the point when the common wonders that are the essential asset of energy are continuous and not drained by human demands [104]. Non-renewable resources are those that are fundamentally exhausted by human use and that would not recoup their potential essentially during human lifetimes such as coal, which does not form normally at a rate that would uphold human use. Likewise, sustainable energy by and large demonstrates a type of energy which is viewed as sustainable, implying that the utilization of alike energy can possibly be maintained well into the future devoid of generating unsafe consequences for upcoming generations. Sustainable energy resources are frequently viewed comprising every sustainable resource, for example, wave power, geothermal energy, artificial photosynthesis, bio fuels, tidal power, hydroelectricity, solar energy, and wind energy [105,106]. It for the most part additionally incorporates advances that improve energy productivity, for example, energy generation, carbon the executives, energy stockpiling and conveyance, and energy utilization and usage with upgraded energy conservation, efficiency, and ecological management.

There are numerous ways to generate electricity, with the age blend as of now ruled by natural gas, coal, and nuclear resources [107]. The worldwide utilization of power catches the achievement and the dispute of energy. In view of the fact that Edison, Tesla, and Westinghouse introduced the main crude electricity networks 130 years ago, electrical innovation has experienced numerous upheavals. Initial use of energy particularly for lighting and electricity is now symbolizing the modern life, communication, entertainment, industry, etc. In the earlier century, 75% of the world has accessed this most adaptable energy carrier [108]. These variations in human lives do not originate from gradual upgrades but from pivotal innovative work on materials that open new skylines. Gigantic openings as of now subsist for progressing from carbon-based energy resources, for example, fuel for motors to electric engines by means of transport, just as from coal-terminated electric power production to inexhaustible, clean solar, atomic, and wind energy sources for electricity, and along these lines significantly expanding the limit and unwavering quality of urban frameworks in high density [109–111]. These progress will need new invention of advanced materials, for example, battery/super capacitor materials for gigantic electrical energy storage, high effectiveness and low-cost solar cells, utilization-safe amalgams for high-temperature power conversion, solid and trivial compounds for turbine blades, superconducting power circulation links, just as efficient materials for highly developed electronics. Furthermore, progressive headways in materials, together with trivial aviation composites, high-temperature motor materials, and propelled composites, have been a basic piece of improving the ability, security, and energy proficiency of transportation medium. Consequently, the association is clear between material research and the energy advancements that human beings depend on today and those requirements for future. Material research covers a wide set of science and engineering disciplines, and looks to comprehend major physical and chemical properties, and after that utilization that comprehension to strengthen the innovation base that people rely on to address the issues for energy, information technology and telecommunications, national security and defense, consumer products, healthcare, and more [112].

1.1.6 Role of Energy Materials in Energy System

All energy advancements need materials for development of advanced energy system, be that as it may, the sorts and measures of materials spent very extensively [113–115]. Material science and engineering is just a single viewpoint of the reaction to the energy challenges; it basically has a significant part to play in generating the propelled energy systems [116]. Formerly, it has contributed essentially to progresses in the protected, dependable, and proficient utilization of energy and accessible common assets. With the appearance of nanomaterials and pioneering multi-purpose materials, material research is essential to assume an expanding job in feasible innovations for harvesting, savings, storage, conversion, and usages [117]. Principal areas of superior material advancement are incorporated yet not restricted to solar cells, batteries and super capacitors, fuel cells, more efficient lighting, thermoelectrics (TEs), hydrogen, superconductors, and advanced energy-harvesting technologies. In nearly all of these regions, gradual upgrades of current advancements are not adequate to address the basic problems regarding robustness, efficiency, and costs. New material research opportunities are, in this manner, expected to configure, expound, and coordinate materials for energy applications. Nanotechnologies and demonstrating activities have been instrumental in this regard, for instance, in the advancement of new electro catalysts for fuel-cell layer anode assembly and in the advancement of new materials for solid-state storage of hydrogen [118,119]. New energy-efficient devices likewise call for superior dispensation advancements for material amplification and incorporation. This is the case for superconductor tapes produced with complex thin-film architectures where there is an ought to control each progression of the manufacture. The final product is a tape of more than a few hundred meters length with a nanostructured superconducting dynamic layer of 1–3 μm thickness. The modern era runs on energy that, in the main, was confined millions of years ago from the sun by photosynthesis in trees and other primeval plants. Today, by consuming coal, oil, and gaseous petrol, this vitality is utilized to fuel transport systems, power matrices, industry, and agriculture, and to warm and cool homes, work, and amusement places. It is clear and well known that fossil energy sources are limited [120]. Moreover, returning enormous amounts of carbon dioxide to the climate expands the amount of energy that it assimilates from the sun with ramifications for the parity of the planet's atmosphere frameworks [121]. It is necessary to create elective wellsprings of energy and to guarantee that energy, regardless of its source, is utilized as productively as could be allowed, which at that point brings about that energy being put away, conveyed, and changed over into heat, light, or movement with insignificant misfortunes. Material research is at the core of advancements for modern energy frameworks. Photovoltaic boards transform sunlight into electricity that can be utilized locally or nourished into the network [122,123]. The most well-known material at present utilized is bulk silicon, in spite of the fact that there are overall endeavors to create all the more exceptionally performing systems based on other semiconductors, thin films, dyes, etc. Fuel cells have the need of a fuel, for example, hydrogen, alcohol, or hydrocarbons that are oxidized electrochemically to generate electricity, which would then be able to be utilized to control a vehicle with a by and large improved effectiveness as contrasted with consuming the fuel in an interior ignition motor [124]. The fuel is oxidized at the anode and is isolated from the oxidant at the cathode by an electrolyte that enables particles to stream inside the electrodes [125,126]. The behavior of the cell relies essentially on these segments and numerous latest materials are being structured for ideal expense and efficiency. In the event that hydrogen is utilized as a fuel, a way to store it on a vehicle is required. Hydrogen-storage materials, in light of light components that dilemma and discharge hydrogen under gentle circumstances are effectively looked for. So also, materials that can absorb carbon dioxide or different toxins can be utilized to help tidy up the results of burning from a fossil fuel power plant, conceivably covered with the goal that it does not enter the atmosphere.

Material research on energy-related materials is currently executing from structural materials and efficient materials to high-energy photons. Different characterization techniques, for instance diffraction and spectroscopy, can be utilized to disclose the atomic structure, composition, and crystalline state of materials, and imaging permits deficiencies and composites to be ascertained. In situ investigations can be the most effective to follow materials or even total devices under working conditions to lookout into the advancement of the mechanism followed by material, for example, lithium-ion battery during the electrochemical cycle [127–130]. From such perceptions, a better comprehension of the performance can be attained and by this means developments and current structures imagined. A noteworthy enhancement in limits with especially an augmentation into the nanoworld has turned out to be always significant for the improvement of superior materials. For instance, the base materials can be thin films, composites, elastomers, high-temperature alloys, polymers, powder metals, and others, while the related devices can be sensors and controllers, optics, and so forth utilized in the creation of the different sorts of sustainable energy systems [131]. Energy is a zone where material innovation will have an especially noteworthy responsibility in gathering the requirements of future [132]. The developing significance of ecological issues is with the end goal that energy age, preservation, storage, and security of inventory will keep on being significant drivers for material innovation. Sustainable energy generation and use are required while simultaneously meeting financial and natural focuses. The high need of energy makes it imperative to continue research, advancement, and displaying of materials for energy applications; it ought to be decided to recoup, catch, and build up the learning base of high-uprightness basic materials for future power world; transferable material arrangements and techniques over the total energy portfolio ought to be analyzed to accomplish most extreme proficiency and upper hands; the worldwide spotlight on energy is a chance to create world-class information and ability in materials for energy production, low energy handling and energy preservation, and to broaden commercial field.

1.2 Energy-Harvesting Materials

1.2.1 TE Materials

The normal future Internet of things (IoT) society may need around a few million sensors for pervasive sensor systems, etc. There is an extraordinary must be developed advancements that can control these sensors without the requirement to supplant batteries. Many attempts have been simulated in producing energy-harvesting materials that can powerfully gather different types of energies from the earth and transform those to electricity. TEs, piezoelectric, magneto electrics, which use thermal energy, mechanical vibrations, and electromagnetic waves are some examples of such types of technologies. TEs use the Seebeck effect for solid-state transformation of heat to electricity [133]. A gauge demonstrating the behavior of TE materials was determined by the figure of merit, $ZT = S^2 \sigma T/\kappa$, where S stands for the Seebeck coefficient, κ for thermal conductivity, σ for electrical conductivity, and T for temperature. The higher the ZT value is, the closer the most extreme change effectiveness approaches the perfect Carnot effectiveness. The normal exchange off between the Seebeck coefficient and electrical conductivity, and rather inconsistent necessity of a material conducting electricity be that as it may, as meager heat conductivity as could be expected consequently, have blocked the improvement of ZT. In this way, different novel standards and materials are being effectively created to develop ZT and the overall thermometric behavior. The materials and applications close to room temperature are particularly expected to be helpful for energy harvesting [134,135]. One unmistakable application is to attempt to utilize body heat by wearable TE modules to control cell phones and sensors [136,137]. Despite the fact that the electrical power prerequisites for such gadgets and sensors are not so high, for instance, as it were 0.1 mW for some electronic following labels. There are numerous attempts proceeding to create devices which need minimum power, similar to spintronic devices. Be that as it may, the effective temperature contrasts which can be used from body warmth is little; what's more, in this way, the TE module ought to be combined with a battery, or further improvement of TE execution should be figured out. As TE energy harvesting can be persistent, even a 10 μW generator can power up a 100 mW class IoT device that uses a battery and faculties or transmits for 1 s in each 3 h. Regardless, the superior the TE behavior, the more extensive are control generation uses of TEs. A further run of the mill necessity for wearable applications is for the material/power generation modules to be adaptable, and in that capacity there is an incredible action continuous specifically to create natural TE materials for this end. In this section, two types of TEs are explained which are listed below.

    i. Organic–inorganic hybrid TEs.
    ii. TE inorganic film: stretchy TE device and micro TE generator.

1.2.1.1 Organic–Inorganic Hybrid TEs

Organic–inorganic hybrid TEs are a captivating field for both organic and inorganic materials scientists. Despite the fact that Shirakawa et al. revealed that the conductive organic polymers [138] most stable organic TE materials are p-type materials and they have lower ZT than inorganic TE materials. Hybridizing with inorganic TE materials is a further way to improve the TE behaviors and an appropriate way to obtain n-type properties. For inorganic TE materials, hybridizing with organic materials offers flexibility and low thermal conductivity: benefits in energy harvesting. For these reasons, organic–inorganic hybrids TEs have been vigorously examined. In recent times, there have been noteworthy investigations on composite TE materials in general. Primarily for some inorganic structures, enhancements of the power factors through instruments like energy separating or regulation doping have been anticipated. There has been additional experiential improvement for highly conducting metallic networks doped into ceramic borides. To procure an easy common image of organic–inorganic hybrid TEs exclusive of any such exotic effects, first expect a specific size of spherical inorganic TE particles blended in with natural materials. It tends to be assumed that we can get the delivered TE power from the inorganic TE particles when the inorganic TE particles network for electric conduction. Here, we can categorize the volume ratio of inorganic TE particles into three regions: (i) non-networking region for the low-volume ratio, (ii) networking region for the middle-volume ratio, and (iii) reverse region for the high-volume ratio. In the non-network region, it can be accepted that the inorganic TE particles do not help out the TE performances. On the contrary, to the extent that we suppose spherical particles, the reverse region is out of curiosity as the inorganic TE materials do not attain elasticity. Therefore, the hybrid effect ought to be seen in the network-forming district. The most extreme proportion for the network-forming district is around 74% where the spherical particles give the packing accomplished with the organic materials. In association with the relation amid the volume ratio and the packing style, the body-centered cubic packing, the primitive cubic packing, and the diamond cubic packing enhance at the ratio of 68%, 52%, and 34%, respectively. It recommends that the inorganic TE particles have the potential to generation produced power in the extensive volume ratio from 34% to 74%. To maximize the TE behaviors, we necessitate deciding the suitable particle sizes to develop the mean free paths of electron and phonon since the particle network for electric conduction can also conduct heat. In comparative system, for instance, percolation theory explains electric and thermal conductions, electrically conductive adhesives; metal-resin composites, and thermal conductive sheets; filler-silicone composites where the interfacial resistance assume the main job [139–141].

Bi_2Te_3-type materials are inorganic TE materials by and large perceived to show high ZT close to room temperature. Identified with the interest in conceivable quantized impacts on the Seebeck coefficient S. 5-nm, 20-nm, and 2-μm Bi_2Te_3 particles have been hybridized with polyaniline [142]. Though the smallest particles stated to illustrate improvement of S, from 17.9 to 120 μV/K, electric conductivity σ maintains or decreases from 37 to 2 S/cm, conditional on the combination condition. This backs the possibility of the significance to control the interface for electric conduction. To explain the difficulty, various additives and high-temperature annealing with thermal-resistive additives have been inspected to attain corresponding TE behaviors with bulk Bi_2Te_3 by improving the drawback of low σ by abridged thermal conductivity [143–145]. As an alternative of surface-sensitive TE materials, carbon nanotubes have also been examined vigorously in composites, since carbon nanotubes encompass self-passivated surfaces to supposedly accomplish 2 orders of magnitude higher TE performances than just the original organic materials. Essentially, the above outcomes can be talked about from a large-scale perspective dependent on permeation hypothesis [146].

One more attractive advance for organic–inorganic hybrid TEs is to exploit molecular intercalation. Wan et al. found that intercalating organic particles brought about making n-type inorganic TiS_2 adaptable and less thermally conductive [147,148]. Furthermore, the carrier mobility is pretentious by the dielectric constant of the intercalation molecules [149]. Consequently, the ZT value for this n-type supple material reaches 0.33 at 413 K, which is considerably greater than the original TiS_2 single crystal. Modules with organic p-type and utilizing this sort of TiS_2 hybrid material as the n-type have additionally been developed with generally high force densities. In general, the hybrids TEs are opening a noteworthy possibility in particular for n-type TE materials in the direction of energy harvesting.

In recent times, additional attempts were made to manage and exploit the interfaces of organic and inorganic materials for improving TE characteristics. Chen and co-workers concentrated on the part of surfactants during the period of the mixing procedure [150]. They point out that in numerous distinctive cases TE improvement in organic–inorganic hybrid materials have not been shown, and quality that to phase division and heterogeneous mixing in the compound. Where the smallest amount of necessity is to have sufficient volume of inorganic material to generate effectual relations. Nevertheless, also by taking into account enough amounts of inorganic materials, if the mixing is poor, consequential in in-homogeneity, then reimbursement from the hybridization will not be comprehended. Different kinds and convergences of surfactants were distorted to thoroughly change the homogeneity of the mixture of PEDOT:PSS and $Bi_{0.5}Sb_{1.5}Te_3$. It was revealed that $C_{14}H_{22}O(C_2H_4O)_n$ with $n=9$ or 10 (TX-100), which has a hydrophobic tail, is more efficient in generating a randomly assigned mixing than frequently used dimethyl sulfoxide (DMSO), foremost for an improvement for these two Seebeck coefficient and electrical conductivity. This outcome highlights the significance to accomplish a homogeneous mixing in this kind of hybrid material [150].

In observe to the excellence of the interface amid the organic and inorganic materials, a thorough theoretical examination was executed by Malen et al. on the TE conveys of a junction of a single organic molecule with inorganic contacts [151]. According to them, the sharp peaks in the density of states at the Fermi level are analogous to the idyllic ZT case of the electronic transport in the course of a single energy level established by Mahan and Sofo. The energy-filtering effect is also well documented and explained to improve TE properties, for example, in poly (3-hexylthiophene) (P_3HT) where nano-composites were created with Bi_2Te_3 nanowires. The interfacial obstruction height ranges to efficiently disperse less-energy carriers below greater electrical conductivity is well thought out to be 0.04–0.10 eV. In the energy-filtering system, less-energy carriers are spotted at the interfaces with greater energy carriers, which can bring more heat carrying transversely and thus improving the Seebeck coefficient. P_3HT was activated with $FeCl_3$ to tune the electrical conductivity with intense doping resulting at an interfacial wall altitude below 0.10 eV in the energy band diagram, corresponding to the area where improvement of the power factor of the nano-composite was observed. P_3HT-Bi_2Te_3 nano-composites created by Bi_2Te_3 nanoparticles did not illustrate improvement which was hypothesized to be caused by the less-energy electron wave functions being capable of circumnavigating the nanoparticles.

1.2.1.2 TE Inorganic Film: Stretchy TE Device and Micro TE Generator

Theoretical studies to advance the TE conversion efficiency attributable to quantum impacts showing up in low-dimensional structures, for example, super-lattices, demonstrated by Dresselhaus et al. [152], inspires thin-film research workers to research TE materials. The utilization of quantum-confinement occurrence develops Seebeck coefficient (S) and manages Seebeck coefficient furthermore, electrical conductivity (σ) fairly free. Phonon scattering turns out to be all the more viably instigated by various interfaces, bringing about lower thermal conductivity. The TE properties of super-lattices in view of different materials have been examined. Upgraded Seebeck coefficient and power factor ($S^2\sigma$) are shown utilizing situated Si/Ge super-lattices detailed by Koga et al. [153]. Lessening of lattice thermal conductivity in $PbTe/PbTe_{0.75}Se_{0.25}$ super-lattices was identified by Caylor et al. [154]. There are a few oxides with non-harmful components that are capable for use as TE materials. Ohta et al. showed the improvement of Seebeck coefficient in $SrTiO_3/Sr(Ti,Nb)O_3$ super-lattices [155]. The quantum-confinement impacts have not been affirmed tentatively, aside from perhaps a couple of works. Tian et al. manufactured InAs nanowires by substance vapor deposition; what's more, pragmatic oscillations in the Seebeck coefficient and power factor attending with the gradual conductance increment (Figure 1.2) because of the one-dimensional quantum confinements in InAs nanowires [156].

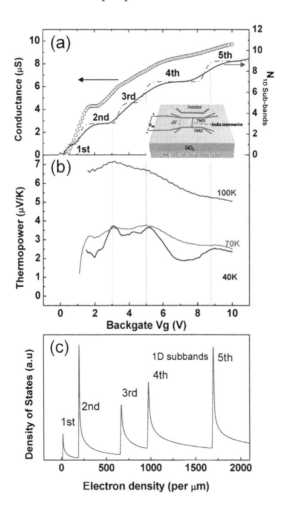

FIGURE 1.2
Gate-tuned conductance and thermopower of InAs nanowire. (a) Comparing the measured conductance (40 K) vs gate voltage (Vg) data (open circle) with the calculated one-dimensional (1D) sub-band occupation with only thermal broadening (dash dotted line) or both the thermal and scattering broadening (solid line) considered, (b) Gate modulation of thermo power (S) at 100, 70, and 40 K. The dashed vertical lines are a guide to the eye, highlighting the appearance of peak in S(Vg) when a 1D sub-band starts to be filled, and (c) Calculated density of states vs. 1D electron density in nanowire with the index of sub-bands marked [157]. (© 2018 The Author(s). Published by Informa UK Limited, trading as Taylor & Francis Group.)

Thin-film TE materials are susceptible to a lot of compact lattice thermal conductivity, and therefore they demonstrate improved TE behavior proportionate to that of bulk materials due to surface and interface scattering of phonons. A research article cautiously examined the effect of particle-size distribution on thermal conductivity in polycrystalline bismuth telluride-based thin films [158]. The lattice thermal conductivity diminishes quickly with the reduction of the particle size, demonstrating the improvement of phonon scattering at the alloys of particles. The entire thermal conductivity of a $Bi_2Te_{2.7}Se_{0.3}$ polycrystalline thin film by means of the mean particle size of 30 µm was 1.6 W/m K. By diminishing the normal size distribution down to 10 nm, the total thermal conductivity accomplishes 0.61 Wm K. In addition, contrasted and TE mass materials, TE gadgets acknowledged by means of thin-film research propose the wonderful benefit of regional and immediate cooling attributable to their ability for scaling down.

In recent times, TE devices constructed using serigraphy stretchy films (Figure 1.3) have been stated by Varghese et al. [159]. Nanocrystal ink of activated $Bi_2Te_{2.8}Se_{0.2}$ was prepared by means of a microwave inspired wet-chemical synthesis technique rooted in low-cost organic solvents and metal salts. The developed nanocrystal ink was acrylic on stretchy polyimide carrier materials. The thickness of the film should be lies in nano range using screen mesh size. The electrical conductivity and Seebeck coefficient of $Bi_2Te_{2.8}Se_{0.2}$ were something like $2–3 \times 104 \ \Omega^{-1}m^{-1}$ and -140 µV/K, respectively, at 473 K. The thermal conductivity was just about 0.6 $Wm^{-1} K^{-1}$, which is extremely low in association with that of a $Bi_2Te_{2.8}Se_{0.2}$ tablet. In this manner, the non-dimensional figure of merit, ZT, accomplished 0.43 at 448 K. As displayed in Figure 1.4, the TE device constructed

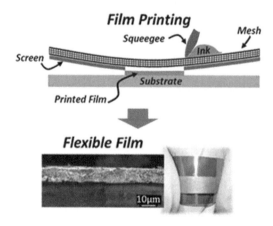

FIGURE 1.3
Diagrammatical representations of the fabrication of stretchy film using serigraphy technique reported by Varghese et al. [157]. (© 2018 The Author(s). Published by Informa UK Limited, trading as Taylor & Francis Group.)

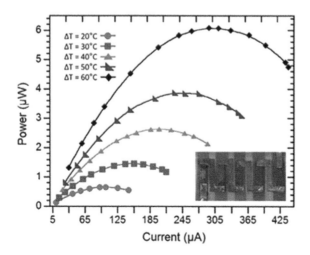

FIGURE 1.4
Current output power description of flexible TE devices. Equipped using serigraphy $Bi_2Te_{2.8}Se_{0.2}$ film. Inset is a picture of the devices [157]. (© 2018 the Author(s). Published by Informa UK Limited, trading as Taylor & Francis Group.)

using a serigraphy stretchy $Bi_2Te_{2.8}Se_{0.2}$ film created a peak performance power of 6.1 μW from a temperature dissimilarity of 60 K.

1.2.2 TE Properties of Lanthanide-Doped Titanates

Non-hazardous and inexpensive TE materials have been considered necessary elongated for extend the consumption of waste heat [160,161]. Transition metal silicides and some oxides were examined as eco-friendly TE materials. Some reduced or donor-doped titanium oxide and sulfide are evidences of n-type metallic performance with a large Seebeck coefficient. The large Seebeck coefficient is accredited to the large effectual mass in the disintegrated Ti3d orbital [162]. They are composed of economically cheap elements and are thermodynamically stable at high temperatures, which make the material potential for wide employment. Nonetheless, as a potential TE material, thermal conductivity was comparatively high. In order to get better figure of merit, several replacements and lattice defects were commenced in favor of diminishing thermal conductivity. Perovskite-type titanate erstwhile premeditated significantly because of its characteristic properties and an assortment of technological applications [163]. The A site is engaged by alkaline-earth and rare-earth elements which can be substituted extensively with each other. In the system of $RETiO_3$ (RE: rare-earth elements), only europium titanate has a cubic structure like that of alkaline-earth titanate [164]. The titanium ion is tetravalent in europium titanate and is not trivalent as in other rare-earth titanates. The structure and some magnetic and electrical properties of europium titanates have been investigated. The assessment of the TE properties of alkaline-earth titanate can be important for the investigation of oxide TE materials.

1.3 Power-Saving Materials

In the past decade, white light-emitting diodes (WLEDs) had attracted greater than ever interest owing to the benefits over conventional incandescent or fluorescent lightings such as high luminous efficiency, long life, energy-saving tendency and eco-friendliness. [165,166]. The industrial approach for WLED mostly depends on the mixture of LED chips with blue/green/blue tricolor materials [167]. At the present time, the industrial and best-known method is the combination of the InGaN blue LED yellow phosphors [168,169]. In the interest of produce the white light with the warm insight parallel to incandescent light, among the best methods is to combine a highly proficient red phosphor giving strong blue absorption with YAG: Ce^{3+} yellow phosphor. At present, the advancement of WLEDs generated by near UV (n-UV)-LED chip excited RGB (red, green, and blue) phosphors are broadly under improvement. Furthermore, field emission displays (FEDs) have also been paid attention as one of the favorable flat-panel displays owing to their high resolution, low power consumption, thin-panel width, self-emission distortion-free images, fast response time, wide-viewing angle, and effective temperature range, in which tricolor phosphors are vital components [170,171]. However, the phosphors used in FEDs are different from those in WLEDs, which have to offer high conductivity to avoid charge accumulation and high stability to tolerate high-energy electron bombardment at low voltage (5 kV) and high current density (10–100 mA/cm²). For enhancing the conductivity of phosphors, reducing the particle size and selecting a conductor as host are two pathways, but most of the researchers are still focused on the previous method and the latter is rarely documented.

1.4 Perovskite Photonics

A considerable lot of the fundamental material characteristics of perovskites that prompted the advancement in photovoltaic efficiencies also facilitate improvements in operation of lighting devices. Perovskites have a keen light-emitting inception, with α, the absorption coefficient, beyond 104 cm⁻¹ near the band edge. An all the more emphatically engrossing semiconductor can proficiently change light over to electrical flow, and, in that order, can uphold higher material gain in a laser. This absorption coefficient exceeds even that of GaAs, a foremost semiconductor in superior business optoelectronic appliance, together with photovoltaic LEDs and lasers.

The electronic and optical operation of solid-state perovskites is exceptionally delicate to the film development technique [172]. Consequent to the revolutionary work on perovskite solar cells, there now exist numerous exceptionally tuned preparing conditions for perovskite thin films [173]. Investigation of the thorough equilibrium of emission and absorption unites brilliant solar cell operation with exceptional LED behavior. The outstanding solar cell functioning of perovskite photovoltaics, in particular their high open-circuit voltage proportionate to their bandgap, envisages proficient function as LEDs. Subsequent to the rush in perovskite photovoltaics administered by progress in thin-film improvement, researchers one more time commenced examining the scenario of these outstanding solar materials in LEDs.

Perovskites, with absolute tunability during the electromagnetic spectrum, are additionally a capable nominee for WLEDs. The first manifestation of perovskite lighting device has been attained using a 2D-layered perovskite with self-caught emissive states and a mixed cation combination of tunable perovskite nanocrystals encompassed in a polymer grid. The essentially low exciton-binding energy in perovskites about a few milli electron volts is a restrictive issue in embryonic highly efficient LEDs. To endorse radiative recombination, thin active regions are essential to imprison carriers three-dimensionally. Manufacturing extremely thin perovskite films with absolute base-material coverage has confirmed complicated and stays a continuous test on the way to superior efficiency. Perovskite nanocrystals own larger exciton-binding energies than their bulk counterparts; nonetheless, preliminary efforts on forming LEDs from these systems have resulted in lower efficiencies than for bulk perovskites, first and foremost as a result of non-ideal surface passivation and poor film configuration. The preeminent performing perovskite nanocrystal LED demonstrated an emission line width of 18 nm with peak EQE near 1% and peak brightness near 2,000 cd m^{-2}, highlighting the initial assurance of these quantum-confined materials for applications in light emission.

Continuous efforts on film construction, customizing electrodes for proficient inoculation, and the improvement of new perovskite energetic areas planned to encourage radiative reconsolidation will persist to conduct the advancement in LED efficiency and encourage higher intensity in feriorinoculation currents.

2

Energy Materials

The worldwide move of energy assembly from petroleum products to sustainable power sources necessitates more effective and dependable electrochemical energy storage devices [174,175]. Specifically, the advancement of electric or hand hydrogen-controlled vehicles calls for a lot better batteries, super capacitors and fuel cells which are right now accessible. Energy is basic to life. Both engineering processing and development of humankind are subject to transformation and usage of energy. Adequate and maintainable creation of energy is classified among the peak significant matter that our general public facade on its approaching the future [176]. Previously, over 80% of the energy for human performance has been gotten from petroleum derivatives like oil, coal and flammable gas, none of which is sustainable. Energy supply ought to be moved to renewable sources previous to the petroleum derivatives which are totally devoured. In reality, the reduction of the Earth's aggregate nonsustainable power sources has started to be noticeable, resulting energy deficiencies and urged upwardly the cost of energy use. Utilizing electrochemical energy storage devices would permit storage of the overflow energy when power supply is beyond request and arrival of it when request gets more than the supply [177–179]. This would make the network increasingly effective, dependable and conceivably less expensive. Lattice scale energy storage necessitates electrochemical appliances with required energy and power yield qualities, long lifetime, high proficiency and ease. Conversely, electrochemical energy storage and alteration appliances are right now effectively followed for impelling electrically driven vehicles, with an objective of expending less oil and producing less ozone-depleting substance in transportation [180–182]. In order to do these, electrically driven vehicles surmount the automobile market. As a matter of first importance, the electrochemical devices must be protected. Furthermore, the energy density of the gadgets should be sufficiently high to give a practically identical driving reach to that of a reservoir of gas or diesel. The appliances ought to likewise be solid and modest sufficient to maintain the deal expenditure and support price of the electrically driven vehicles which are focused. It is practically difficult to discover one sort of device that satisfies each of the necessities presented by various down-to-earth applications. A potential arrangement is to create different sorts of electrochemical energy storage and change devices with various energy and control capacities and join them in innovative approaches. Ferroelectric materials together with normal ferroelectrics,

DOI: 10.1201/9781003381907-2

relaxor ferroelectrics and anti-ferroelectrics illustrate the impact of uncon-
strained electric polarization which can be turned around by electric field
application [183,184]. On account of their electric polarization behaviors and
unique dielectric properties, they have been generally examined for different
applications, for example, capacitors, energy harvesting and storage, high-
power electronic transducers, electrocaloric cooling/heating devices, microwave hardware and nonvolatile memories [185,186]. For such utilizations of
dielectric or ferroelectric materials, low-spillage flow and high breakdown
quality notwithstanding high electric polarization and dielectric consistent are required. While epitaxial ferroelectric films have higher densities
and lower defect concentrations than mass and properties can be improved
by epitaxial requirements [185,187,188], making them proficient of surviv-
ing high electric fields and possessing exceptionally large volume specific
energy storage density [189]. They are still far from idyllic. Especially, oxide
thin films have crystallographic defects which unenthusiastically impact
breakdown strength [190]. The enduring drive toward neatness of electronic
circuits and devices is inspiring the progress of new 'thin film' materials
[191]. The group of perovskites with the general formula ABO_3 (A: cation at
the vertices of the 3D cube, B: cation at the body center) is a significant class
of ferroelectric materials. Industry-standard ferroelectric perovskites contain
lead, which is lethal and ecologically unpleasant on the grounds that sans
lead materials typically showed second rate ferroelectric properties to those
of the toxic materials. Be that as it may, in light of the developing demand for
green materials with limited effects on well-being and condition, sans lead
syntheses are earnestly required [192]. Solid solutions capitulate the opportunity to generate superior, surprising or better properties ahead of averaging
of their end member properties. Careful choice of end members and their
compositions is required. Although ferroelectric epitaxial solid solution thin
films have been reported so far [193,194]. The idea of enhancing ferroelectric
and dielectric properties in Pb-free ferroelectrics films by selecting the right
chemistry for defect reduction at the same time as enhancement of the temperature of the dielectric maximum has not been considered to date [195].

2.1 Role of Activators

Lately, ecological hazardness has become an essential issue around the world
[196]. Incidentally, photocatalysis has been proposed as a naturally agreeable
methodology and a hopeful technique to expel contaminant from water and
air because of the conceivable oxidation of natural contaminations to carbon dioxide and water [197,198]. TiO_2 is the most recently exposed and more
extensively examined photocatalyst because of its higher photo activity, comparatively inexpensive, low-insalubrity and excellent chemical and thermal

constancy [199–201]. Regrettably, titanium dioxide is responsive only to UV light by reason of its enormous band gap, and it has less quantum efficiency, ensuing from the quick reconsolidation speed of photo-generated electron–hole pairs [202]. Different methodologies of transforming TiO_2 have been useful to survive the previously mentioned disadvantages, for example, metal and nonmetal doping, surface alteration by dyes, dignified metal nanoparticles or narrow band-gap semiconductor elements [203–205]. Between these schemes, the research on TiO_2 tailored with lanthanides has all the earmarks of being exceptionally effective technique to tune the reaction of the semiconductor to the apparent light region and to upgrade its photocatalytic properties [206–208]. The ebb and reviews on rare earth metal-activated TiO_2 center around the synthesis technique and characterization of the photocatalysts, the photocatalytic debasement of organic molecules and the photosplitting of water in the company of RE-TiO_2, La, Ce, Eu and Gd-TiO_2, the use of Gd-TiO_2 for photoenergy applications [209], the employment of RE-TiO_2 for the photocatalytic restoration of effluent [207], visible-to-ultraviolet upconversion procedures [210] and slim RE-TiO_2 films formed by spray deposition [211,212]. Nonetheless, it is difficult to assess yet how rare earth ions make influence over visible light reaction of TiO_2-based photocatalysts. In perspective on this, this chapter covers significant features accompanying the photocatalytic activity system over preferred RE-TiO_2 photocatalysts and the bond among the method and the anti-Stokes upconversion procedure. Furthermore, a computational description of the electronic structure and optical properties is given to propose potential thoughts to specialists in this field for the plan of new materials upheld by computational techniques. The synthesis strategies and surface properties of RE-TiO_2 are additionally talked about exhaustively.

2.2 Unique Properties of Lanthanides

Rare earth metals have shown extraordinary potential as TiO_2 dopants in the red shift of absorption as well as in expanding the temperature of the anatase to dispersibility change and in the capacity to frame buildings with different Lewis bases, for example, alcohols, aldehydes, acids, amines and thiols [213–215]. Moreover, TiO_2 materials tailored by RE^{3+} ions typically exhibit luminescent properties. Rare earth ions with ladder-like energy levels entrenched in a suitable inorganic host lattice can emit UV or visible light in the course of the chronological absorption of multiple near-infrared photons [216]. This cycle, which can bring about the transformation of light from the close infrared and visible spectral range to the ultraviolet wavelengths, could be utilized to animate wide band semiconductor, like TiO_2 [217]. The lanthanides comprise of 15 components beginning with atomic number 57 (lanthanum) to 71 (lutetium). Scandium and yttrium are artificially comparable; along these lines, at long last 17 components establish the rare earth series. The glow of rare earth particles emerges from the f–f electronic transitions inside their half-filled 4f orbitals. These orbitals are sarcastically protected from the encompassing microenvironment by the filled 5s and 5p orbitals, implying that there are practically no annoyances of these transitions. Subsequently, these transitions appear as several narrow emission bands for the transmitting lanthanide particle. This phenomenon gives the lanthanides exceptional compound properties in photocatalytic applications. Just four primary lanthanide particles can hypothetically actuate the TiO_2 photocatalyst through the Vis-to-UV or NIR-to-UV upconversion measure, specifically Er^{3+}, Ho^{3+}, Nd^{3+} and Tm^{3+}. The upconversion cycle can be accomplished through the chains of the ground state absorption (GSA) and excited state absorption (ESA) (Figure 2.1) [218]. Through the anti-Stokes procedure,

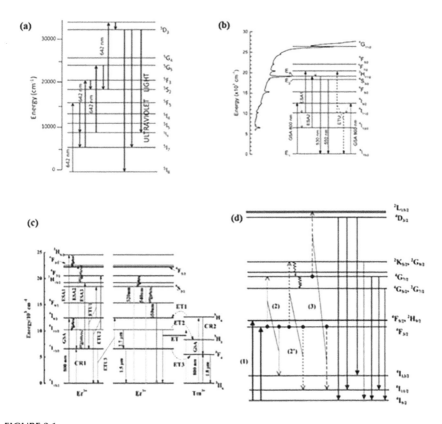

FIGURE 2.1
Schematic energy level diagrams of an excitation path for the upconversion emission of ultraviolet light: a) Ho^{3+} under 642 nm excitation, b) Er^{3+} under 522 nm excitation, c) Er^{3+} and Tm^{3+} when pumped at 800 nm and d) Nd^{3+} under 578 nm excitation; GSA – ground state absorption, ESA – excited state absorption. (Reprinted with the permission from Ref. [220–223]. Copyright © Elsevier publications.)

at least two photons are consumed consecutively by a material to arrive at an excited state, and one higher-energy photon can be unconfined [219]. Therefore, just the TiO_2 doped by Er, Ho, Nd and Tm was depicted in the review and an endeavor was made to connect action of these photocatalysts with their upconversion properties.

2.3 Preparation Methods and Surface Properties of RE-TiO_2

2.3.1 Erbium-TiO_2

2.3.1.1 Morphological Study of TiO_2:Er^{3+}

Usually, Er-TiO_2 conceivably acquired through hydrothermal/solvothermal synthesis route and sol–gel preparation technique. Previously published articles have uncovered that the hydro/solvothermal and sol–gel strategies are still the absolute most normally utilized preparation techniques for the synthesis of Er-TiO_2 [224–226]. These kinds of techniques have turn out to be ever more popular as easy, commercial synthesis procedure to generate Er-TiO_2 with superior purity at a comparatively minimum temperature. Altogether outcomes, Er-TiO_2 photocatalysts became working by means of different morphological behavior together with zero-, one- or two-dimensional structures (Figure 2.2). This morphological behavior determines the material and structural properties of RE-TiO_2 in addition to its luminescence and photocatalytic properties. To synthesize Er-TiO_2, titanium (IV) isopropoxide, titanium (IV) butoxide, TiO_2 colloidal solution perhaps utilized as TiO_2 precursors, while $Er(NO_3)_3$·xH_2O, $ErCl_3$ and Er_2O_3, could act as erbium precursors [227–232].

  Among the entire material preparation methods, spin coating [233], implantation [234], magnetron sputtering [235], and anodic [236] and thermal plasma oxidation [237] are occasionally used. All these methods generate Er-TiO_2 photocatalysts like thin layer. Furthermore, the electrochemical oxidation of Ti foil within the sight of $Er(NO_3)_3$ in the ammonium fluoride/ethylene glycol natural electrolyte permits arrangement of situated nanotube clusters altered with erbium particles [238]. Moreover, bulk of this doping approaches have no control on the structural phenomena of the acquired Er-TiO_2; therefore, a straight assessment of the photoactivity among the pristine and tailored samples is consistent. However, in few cases, the structure of the TiO_2 photocatalyst seems to be somewhat affected by the existence of Er ions. It was tracked down that the synthesis route they applied for the readiness of Er-TiO_2 photocatalyst produces a homogeneous dissemination of round particles with sizes under 15 nm [239], be that as it may, as the Er^{3+} content builds, the normal size continuously diminishes, arriving at a worth of 10 nm for 4 at.% Er^{3+} [240].

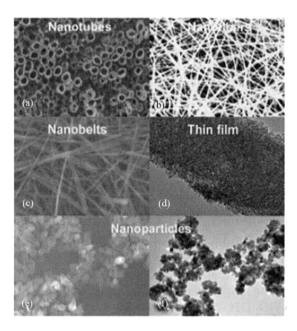

FIGURE 2.2
(a–c) SEM and (d–f) TEM images of the Er-TiO$_2$ morphologies. (Reprinted with the permission from Ref. [241–246] Copyright © Elsevier publications.)

2.3.1.2 Crystal Structure of Er-TiO$_2$

It must be noted that classified among main considerations that significantly impact the properties of the tailored photo catalysts is the crystal structure, in particular the crystallite size and unit cell parameters. Yang et al. [245] synthesized Er- activated TiO$_2$ nanofibrous films via an electrospinning technique. Furthermore, they saw that the crystallite size diminished from 17.9 to 8.1 nm as the erbium content commenced into TiO$_2$ enhanced from 0 to 1.5 mol%. They likewise tracked down that the cell parameters of Er-TiO$_2$ were like pure TiO$_2$. These outcomes exhibited that Er^{3+} species exist at the crystal boundary instead of the internal crystalline structure of TiO$_2$ because the replacement of a little Ti particle (68 pm) by an enormous Er particle (100 pm) leads to an expansion in the lattice parameters and a shift in the diffraction patterns [245]. A comparative impact was found in crystallize size lessened from 30 to 16 nm after erbium doping, as the unit cell volume for the Er-TiO$_2$ test was near to that of the pure anatase TiO$_2$. The experiential changes were authorized to the presence of lanthanide particles in the TiO$_2$ host, which pilot to strain and stress in the lattice [247]. Highlight the fact that Er-doped TiO$_2$ nanotubes are synthesized and seen that the diffraction peaks of the activated photocatalyst moved to one side. In view of Bragg's law, they presumed that the left interpretation of the diffraction peaks showed that the

Er^{3+} was fruitfully commenced into the lattice of the TiO_2 nanotubes [244]. An additional significant point that ought to be referenced is the impact of erbium doping on the impediment of the phase transition of dispersibility to rutile. A few publications have shown this occurrence, when erbium ions were doped into TiO_2 lattice [245,248]. It was seen that doping by erbium particles repressed the presence of the dispersibility stage when calcined at same temperature, that is, 500°C. It was suggested that this hindrance may happen because of the adjustment of the anatase stage by the encompassing rare earth components through the arrangement of Ti–O-rare earth component bonds. The connection between the diverse tetrahedral Ti atoms or between the tetrahedral Ti and the octahedral Ti prohibits the phase transformation into dispersibility [245].

2.3.1.3 BET Surface Area and Optical Properties of Er-TiO$_2$

Concerning the BET surface area, Er-TiO_2 photocatalysts demonstrate comparatively noble values; in all instances, the surface area was somewhat exceeding that of immaculate TiO_2. It was seen that the BET surface region logically expanded as the amount of erbium expanded from 102 to 116 m^2/g [248]. Simultaneously, it was seen that the center shows a thin allocation, with a normal center of 9–12 nm with superior normal qualities as the amount of erbium was greater than before [249]. Rather than the BET surface area, the writing information didn't give clear data about the band structure of erbium-doped TiO_2 photocatalysts. A few researchers have noticed a decline in the band gap of Er-TiO_2, in spite of the fact that different researchers have discovered that the intended band gap of Er-TiO_2 materialized to be similar to that of immaculate TiO_2. Consequently, the presence of erbium could conceivably influence the band structure of the anatase. For instance, Mazierski et al. [238] and Obregón et al. [243] discovered that activation by erbium didn't outstandingly influence the absorption edge of TiO_2. Moreover, they noticed a little and reformist blue shift. A comparative impact was additionally seen by Yang et al. [245] wherein Er-TiO_2 photocatalysts were synthesized by an electrospinning method. These outcomes are inverse to those originated by Castaneda-Contreras et al. [250] for Er-TiO_2 structure synthesized by the sol–gel synthesis technique. They proposed that the consolidation of erbium into the structure instigated a checked red shift and the ensuing narrowing of the band-gap values. The development in the photoactivity was for the most part ascribed to the lower band-gap values and, to lesser degree, to energy transfer from Er^{3+} to TiO_2. Similar outcomes to that of Castaneda- Contreras et al. [250] were observed by Lee et al. [251]; that is, the energy gap values were 3.25, 2.85, 2.85, 2.85, 2.81 and 2.89 eV for 0, 0.1, 0.3, 0.5, 0.7 and 1.0 mol% of Er, respectively. Ultimately, the writing information is not predictable, and this angle ought to be explained later on. Moreover, for erbium-doped photocatalysts, various lines show up in the scope of 400–700 nm contrasted with the TiO_2 absorption spectra, and those lines can be related with the excitation of the erbium species. The absorption lines situated at around 489, 520 and 653 nm alongside a little tail at 800 nm relate to the transitions from the Er^{3+} ground state $^4I_{15/2}$ to the higher vitality levels of $^4F_{7/2}$, $^4H_{11/2}$, $^4F_{9/2}$ and $^4G_{9/2}$, respectively [252].

2.3.1.4 Upconversion Properties of Er-TiO$_2$

The upconversion properties of the Er-TiO$_2$ photocatalysts have regularly been contemplated. On account of NIR irradiation, the nearness of erbium seems to advance the upconversion procedure, siphoning photons into the UV run in the TiO$_2$ structure. It was accounted for that Er-TiO$_2$ show a feeble UV photoluminescence (PL) emission at around 390 to 410 nm after excitation at 980 nm [253]. This upconversion process includes consecutive three-photon absorption. At that point, by multiphoton relaxation, the $^2G_{7/2}$ excited state rots to the lower $^2G_{11/2}$ and $^2H_{9/2}$ states. The photoluminescence emission in the UV is then delivered by the $^2G_{11/2} \rightarrow {}^4I_{15/2}$ transition, giving a weak emission at around 390–400 nm [254]. Thus, the progress of the photocatalytic proficiency may be identified with the expanding number of accessible photons with a suitable energy. It was realized that by exciting the Er-TiO$_2$ photocatalyst with 980 nm light, PL spectra in the UV region show that two weak emission bands reside almost 390/3.18 and 415/4.14 nm/eV [255]. Rather than these outcomes, just a 490 nm green emission and a 670 nm red emission are subsequent to excitation of the Er-TiO$_2$ thin film by a laser of 980 nm [256]. The upconversion properties of the Er-TiO$_2$ samples in the form of planar waveguides that acquired via sol–gel method accompanied by dip-coating were deliberated by Bahtat et al. [257]. In these circumstances, four broad bands were witnessed approaching around 410, 525, 548 and 660 nm. Nonetheless, Salhi and Deschanvres [258] achieved RE-TiO$_2$ nanopowders, which showed strong, weak and very weak emission bands reside 550, 525 and 655 nm, respectively. In outline, an extremely delicate emission band situated at roughly 390 nm affirmed the upconversion procedure proposed, thus erbium changed over the NIR photons to high-energy UV photons.

2.3.1.5 Photocatalytic Properties of Er-TiO$_2$

The photocatalytic activity of the Er-TiO$_2$ photocatalysts was researched under UV–Vis, Vis and NIR irradiation utilizing aqueous solutions of methylene blue (MB), phenol, methyl orange, rhodamine B and orange I as model pollutants. These model responses have been utilized to examine both the impacts of the erbium content fused into the TiO$_2$ structure just as the calcination temperature on the photoactivity. Regardless, photocatalytic action has been evaluated utilizing distinctive arrangement and irradiation sources, for example, 500 W xenon light, 200 W Hg-Xe light, deuterium-tungsten light, 160 W high-pressure mercury light, 8 W medium-pressure mercury light and light-emitting diodes (LEDs). The impacts of the erbium content fused into the TiO_2 structure and the calcination temperature on the photoactivity under sun-powered reenacted light utilizing MB as the model toxin [259]. It was demonstrated that the photocatalytic action of the erbium-doped TiO_2 was exceeding that of immaculate TiO_2 under simulated solar light [259,260]. The progress of the 4f electrons of Er^{3+} improved the optical adsorption of photocatalysts and profited the division of photo-generated electron–hole pairs. TiO_2 doped with 0.5 mol% of erbium had the most elevated photocatalytic accomplishment, and the photoactivity decreased by expanding the doping measurement. It was expressed that lower photoactivity was because of higher number of recoupling centers in light of the fact that surplus measures acknowledged that RE metals cover the surface of TiO_2. On account of the calcination temperature, it was observed that the photoactivity of the photocatalysts which was synthesized from 400°C to 700°C diminished [260]. Incorporating these outcomes with the crystalline period of TiO_2 demonstrates that the nearness of the dispersibility stage was accountable for the lower photocatalytic action in the decomposition of organic compounds [259,260]. Reszczyńska et al. revealed qualities of Er-TiO_2 photocatalysts acquired by means of sol–gel technique. A high-photocatalytic execution in the phenol photo degeneration under visible light irradiation was noticed for photocatalysts synthesized via the hydrothermal strategy [261]. All RE-TiO_2 tests synthesized by the hydrothermal strategy had higher BET surface domain and lower crystallite size than that prepared by the sol–gel method. Regardless, photocatalysts arranged by the sol–gel technique have superior measures of RE_2O_3 on their surfaces and less OH^- groups and Ti^{3+} moieties than the powders delivered by means of the hydrothermal method [261]. Zheng and Wang also investigated the impact of the erbium content infused into the TiO_2 structure on the photoactivity [262]. They found that the ideal dopant stacking rose to 1 mol%, which is comparable to the most extreme photocatalytic degradation rate [262]. Moreover, the photo degradation of rhodamine B and phenol under LED irradiation ($\lambda = 517$–522 nm) exhibited that Er-TiO_2 photocatalysts can display photocatalytic movement under obvious light further than the absorption edge of TiO_2. The photocatalytic execution of the Er-TiO_2 photocatalysts in the model responses to phenol and MB in the aqueous phase and toluene in the gas phase degradation within the sight of irradiation with various spectral ranges of the light. In this examination, improved photon performances have been experienced for the response under NIR irradiation than under UV surroundings. UV-instigated photoactivity was allocated to electrons catching by Er^{3+} particles filling in as foragers, while photocatalytic action under NIR was credited to the upconversion properties of the Er-TiO_2 photocatalysts, bringing about exciting photons in the UV run into the TiO_2 structure [255]. Castaneda-Contreras et al. [263] investigated Er-TiO_2 photocatalysts synthesized by the sol–gel technique in the response of MB degeneration under visible light. They proposed that fuse of erbium into the structure actuates a marked red shift and an ensuing tapering of the band-gap values. The development of the photoactivity was basically ascribed to the smaller band gap and, to a smaller degree, to energy transfer from Er^{3+} to TiO_2. To assess the photocatalytic action of as-synthesized phosphors, the energy of orange I photocatalytic consumption utilizing Er-TiO_2 photocatalysts has been examined. The acquired outcomes exhibited that the deterioration and calcination of orange I under both UV and illumination were progressively productive within the sight of Er-TiO_2 photocatalysts than within the sight of immaculate TiO_2. The creators reasoned that the higher action upon noticeable light may be ascribed to the change of the 4f electrons of Er^{3+} and the red shift of the optical absorption edge of TiO_2 via doping of erbium ion [263]. Bhethanabotla et al. [264] deliberated Yb/Er-TiO_2 photocatalysts in the response of phenol and Rose Bengal photodegradation under artificial electromagnetic radiation and illumination at different narrow wavelengths. They saw that phenol and Rose Bengal were debased uniquely by illumination utilizing wavelength that approached to 405 nm. The upgraded photoactivity was portrayed to photon energy of the excitation source which is comparative with the band-gap and contrasts in chemistry rather than to upconversion impact. Then again, Pickering et al. [264] in their examination showed that although upconversion diminished reunification rate further band-gap variation, yet expanded impurities adsorption at the outside of photocatalysts was the most significant purpose behind the enhanced functioning of erbium–based-TiO_2.

At last, to improve visible light effect of Er-TiO_2 and thus photo degradation of contaminants, some extra adjustments of these photocatalysts have been applied, like carbon sensitization and Fe-Er co-doping [265], Er-Yb co-doping [266] and composite systems, together with YAG:Yb^{+3}, Er^{+3}/TiO_2 [267], YSO:Pr^{+3}/TiO_2 and YAG:Er^{3+}/TiO_2 [268]. The structural and physical properties of Er-TiO_2 photocatalysts and explanation of photocatalytic system used is listed in Table 2.1.

2.3.2 Holmium-TiO$_2$

The structural, physical and photocatalytic properties of Ho-TiO_2 photocatalysts are outlined in Table 2.2. To synthesize Ho-TiO_2 photocatalysts as zero-dimensional (nanoparticle) and one-dimensional (nanowire) morphologies, two most normal techniques, to be specific, the sol–gel and hydrothermal techniques, separately have been utilized. Concerning morphology of Ho-TiO_2, Ho-TiO_2 nanoparticles had comparatively consistent sphere-shaped particles, allocated in small sizes, and superior dispersability proportionate to pristine TiO_2 nanoparticles [269], demonstrating that holmium doping could develop the molecule morphology and impede the particle development of TiO_2 upon heat treatment. Additionally, on account of photocatalysts, the particle diameter of holmium-doped TiO_2 was lesser than immaculate TiO_2, and Ho-doped TiO_2 exhibited an anomalous sphericality [270]. It was reported that Ho-TiO_2 photocatalysts as nanowires utilized hydrothermal route. They concluded that nanowires with a consistent scattering and with

TABLE 2.1

The Structural and Physical Properties of Er-TiO$_2$ Photocatalysts and Description of Photocatalytic System Used to Estimate Photocatalytic Activity

Sr. No.	Preparation Method/Obtained Structure	Chemical Characteristics of RE Species	BET Surface Area (m²/g)	Band-gap (eV)	Unit Cell Parameter	Photocatalytic Activity Setup	Upconversion Measurement	Ref.
1	Spin coating Thin film	Ion	-	3.25-pristine TiO_2, 2.81-$ErTiO_2$	-	-	-	[248]
2	Hydrothermal Nanoparticles	Ion	102-pristine TiO_2 110-$ErTiO_2$	3.16-pristine TiO_2 3.18-$ErTiO_2$		Model contaminant: Phenol and methylene blue Irradiation source: 200 W Hg-Xe lamp (UV–Vis–IR, UV–Vis, Vis and Vis–IR)		[251]
3	Sol–gel Thin film						λ_{ex} = 800 nm λ_{em} = 410, 525, 548 and 660 nm	[257]
4	Hydrothermal Nanoparticles	Oxidation state	102-pristine TiO_2 110-$ErTiO_2$	3.16-pristine TiO_2 3.18-$ErTiO_2$	136.30-pristine TiO_2 136.35-$ErTiO_2$	Model contaminant: Phenol, toluene and methylene blue Irradiation source: 200 W Hg-Xe lamp (UV–Vis–IR, UV–Vis, Vis and Vis–IR)	λ_{ex} = 980 nm, λ_{em} = 390 and 415 nm	[243]
5	Electrospinning Nanofibrous film	Ion		Blue shift	136.28-pristine TiO_2 136.69-$ErTiO_2$	Model contaminant: methylene blue Irradiation source: 500 W Xe lamp		[245]
6	Sol–gel Nanoparticles	Ion		Red shift		Model contaminant: Orange I Irradiation source: 8 W Hg lamp (UV) and 70 W sodium lamp (Vis)		[263]
7	Thermal plasma oxidation Nanoparticles	New phase $Er_2Ti_2O_7$				Not tested	Not tested	[272]
8	Anodization Nanotubes	Ion			Increased	Model contaminant: Methylene blue Irradiation source: 160 W Hg lamp		[244]

TABLE 2.2

The Structural and Physical Properties of the Ho-TiO$_2$ Photocatalysts and Description of Photocatalytic System Used to Estimate Photocatalytic Activity

Sr. No.	Preparation Method/Obtained Structure	Chemical Characteristics of RE Species	BET Surface Area (m²/g)	Band-gap (eV)	Unit Cell Parameter	Photocatalytic Activity Setup	Upconversion Measurement	Ref
1	Sol-gel Nanoparticles	Ion	–	3.25-pristine TiO$_2$ 2.81-ErTiO$_2$	−0.1360-pristine TiO$_2$ 0.1365-Ho-TiO$_2$	Model contaminant: methyl orange Irradiation source: Mercury lamp	Not tested	[273]
2	Sol-gel Nanoparticles	Ion	38.42-pristine TiO$_2$ 76.76 for -Ho-TiO$_2$	No data	0.1234-pristine TiO$_2$ 0.1245-Ho-TiO$_2$	Model contaminant: methyl orange Irradiation source and range: 500 W high-pressure mercury lamp		[270]
3	Anodization Nanotubes	Oxide		3.35-pristine TiO$_2$ 3.30-ErTiO$_2$	136.64-pristine TiO$_2$ 136.73-Ho-TiO$_2$	Model contaminant: Phenol and toluene Irradiation source and range: 1000 W Xe lamp and LEDs (465 nm)	λ_{ex} = 375, 530, 650 and 980 nm λ_{em} = not observed	[238]
4	Sol-gel Nanoparticles	Ion			0.1225-pristine TiO$_2$ 0.1235-Ho-TiO$_2$	Model contaminant: methyl orange Irradiation source: 500 W high-pressure mercury lamp		[274]
5	Hydrothermal Nanowires	Ion			Model contaminant: methylene blue Irradiation source and range: 175 W high-pressure mercury lamp	Model contaminant: methylene blue Irradiation source and range: 175 W high-pressure mercury lamp		[271]

an external width and length of 15 and 500 nm, respectively, have occurred at 150°C [271].

Concerning the crystal structure, likewise to Er-TiO$_2$, holmium doping efficiently restrains the advance conversion from anatase to rutile. The constraint of the phase conversion conceivably attributed to the consolidation of the anatase phase via neighboring lanthanides during the Ti–O–Ho interface [270,271]. It was likewise seen that the crystallite size of the Ho-TiO$_2$ diminished with an expansion in the doping of holmium amount [270]. Moreover, investigations have indicated that the crystallite size expanded with the expansion of calcination temperature, which inferred that doping of holmium hindered the development of the crystallite size [270]. It tends to be expressed that the BET surface zone additionally expanded on doping of holmium because of the abatement of the crystallite size. A further significant part of Ho-TiO$_2$ photocatalysts is the disquieting of band-gap structure [270]. A little blue shift of the incorporation outlines in the Ho-TiO$_2$ in comparison with immaculate TiO$_2$. This blue shift was experiential by Cai et al. [273] as well and attributed to the quantum size effect since Ho-doping significantly concealed the crystal size of TiO$_2$. On the contrary, a slight red shift of the absorption outline ranges from 300 to 450 nm was experiential in the Ho-TiO$_2$ nanowires by Zhou et al. [271]. These outcomes to the charge move progress amid the f electrons of Ho^{3+} particles and the TiO$_2$ conduction or valence band. The absorption edge shifted to a higher wavelength, which was subject to the measure of Ho^{3+} joined into TiO$_2$ lattice [271]. As indicated by the accessible literature, the photocatalytic properties of the Ho-TiO$_2$ photocatalysts were examined chiefly just upon UV–Vis irradiation utilizing a high-pressure mercury light as an excitation source and utilizing colors as the model toxins (see Table 2.2). In each of the investigations, the photocatalytic activity of the Ho-TiO$_2$ photocatalysts was clearly higher than that of flawless TiO$_2$ [273]. The better concert of Ho-TiO$_2$ has been clarified by the way that the Ho-doped photocatalysts have a smaller crystallite size and can improve the detachment effectiveness and hinder the reunification proficiency of the photo-generated electron–hole pairs [274]. Also, the development in the crystal matrix makes oxygen opportunities, which creates shallow energy states at the base of the conduction band and fill in as electron-trapping sites in TiO$_2$ [273]. In the interim, shallow energy states presented by Ho particles in the top valence band filled in as hole-trapping sites. The partition of the charge carriers was credited to this trapping system. The ideal amount of dopant approached to 0.3 mol% for the greatest photocatalytic degeneration rate when Ho-TiO2 was calcined at 500°C, and the ideal calcination temperature was 600°C when amount of holmium was 0.5 mol% [271,273,274].

2.3.3 Thulium-TiO$_2$

There are merely not many of studies concerning thulium-doped or codoped compounds in diverse properties, even though Tm-TiO$_2$ photocatalysts encompass acknowledged extremely slight investigate attention. Tm-doped

TiO_2 photocatalysts were prepared via low-temperature hydrolysis reaction [275,276] in addition to a ball-milling method [225,277]. The impacts of the annealing temperature plus amount doped of thulium on the structural and electronic applications of the obtained $Tm\text{-}TiO_2$ materials were examined. Anatase was the major phase in the $Tm\text{-}TiO_2$ samples annealed at 500 and 700°C, at the same time rutile was the main phase in the samples calcinated at 900°C [275]. It is significant that it was experiential just merely the phase consequent to TiO_2 for the materials calcinated at 500 and 700°C, while on account of the samples annealed at 900°C, the pyrochlore period of $Tm_2Ti_2O_7$ was also obtained [275]. Tm and Ti were framed on account of the $Tm\text{-}TiO_2$ photocatalyst calcinated at 900°C. Consequently, the Tm^{3+} particle might be consolidated into the anatase structure; however not into the rutile one, most likely in light of the fact that the ionic radius of Tm^{3+} is impressively superior to that of Ti^{4+} and the anatase structure is less thick than the rutile one [103].

For $RE_2Ti_2O_7$ materials, the pyrochlore structure was framed at a temperature higher than 800°C. Consequently, all the above investigations demonstrated that the structure of the Tm-doped TiO_2 nanoparticles was perhaps thermally restrained. Change of titania with thulium prompted an expansion in the BET precise surface zone of the acquired photocatalysts. It is not considered an essential aspect for enhancing the photocatalytic movement of Tm-doped TiO_2 samples [276,278]. Additionally, XPS examination affirmed the existence of Tm^{3+} in the $Tm\text{-}TiO_2$ samples, which prompted the age of basic twists and oxygen opportunities to keep up the nearby lack of bias in the lattice [276]. The presence of thulium (III) in the materials likewise affected the adjustments in the UV–Vis absorption spectra, presumably as a result of the existence of new states in the band gap, which creates another photonic absorption procedure and demonstrates photoluminescence properties [275]. Santos et al. [278] demonstrated that Tm-activated materials offered numerous absorption bands centralized in the region of 466, 685 and 785 nm, which communicated to the transitions of Tm^{3+}. The photocatalytic activity of Tm-activated TiO_2 and pristine TiO_2 nanoparticles was considerable to investigating the reaction of MB degeneration beneath UV light irradiation using an actinic lamp emitting at around 360 nm. Navas et al. [275] examined the photocatalytic efficiency of pristine rutile TiO_2 as well as 2 and 4.3 at.% Tm-doped TiO_2 samples, which are self-possessed of the rutile and anatase TiO_2 in addition to pyrochlore $Tm_2Ti_2O_7$ phases. These investigation outcomes were additionally affirmed by hypothetical counts, which uncovered that the Ti-O-Ti angle in the pyrochlore phase was in the region of 136°C and this property impacted the superior photocatalytic action of the pyrochlore phase since charge carriers can move effectively in the lattice. Critically, the charge thickness and ELF studies show that the presence of the Ti-O-Ti bonds empowered great charge mobility in the pyrochlore phase [275]. The impact of the calcinating temperature of Tm-activated TiO_2 materials on the photocatalytic action. It was intriguing to take note of that materials calcinated both at 500 and 720°C, displayed lower photoactivity than immaculate TiO_2.

In any case, all Tm^{3+}-activated materials calcinated at 900°C displayed a critical development in the photoactivity (up to 94% debasement), 40% more than the pure TiO_2, which was brought about by the arrangement of a blend of anatase, rutile and new pyrochlore $Tm_2Ti_2O_7$ stages [278]. Further investigation shows that the photocatalytic action of the Tm-activated materials calcinated at 900 was broke down, and the response rate expanded by an aspect of 2.35 in the material with 2.0% Tm [276]. This outcome could be attributed to the existence of pyrochlore, which is concurred *i*th past investigations [278]. Table 2.3 shows the structural and physical properties of the various $Tm-TiO_2$ photocatalysts.

2.3.4 Neodymium-TiO$_2$

Dissimilar types of approaches for the synthesis of Nd^{3+}-activated photocatalysts have been projected: sol–gel [279], a microwave-assisted mild temperature route [280], low-temperature synthesis [281] in addition to chemical coprecipitation peptization [282] and hydrothermal methods [283]. Up to this point, the outcome of the amount of Nd on the physical structure and photocatalytic activities of activated materials was thoroughly examined. An expansion in the amount of neodymium in the $Nd-TiO_2$ samples brought about an increment in the crystallite size of the photocatalysts [284]. In another examination, the titania crystallites diminished in size as the content of activated neodymium particles expanded, which was clarified by the adsorption of the neodymium species on the outside of titania, bringing about the restraint of TiO_2 crystallite growth [285].

The differences in the ionic radii keep the neodymium particles from efficiently being integrated into the crystal structure of TiO_2; consequently, neodymium is most likely restricted at the outside of the TiO_2 nanocrystals [286]. It is likewise significant that the insignificant modification in the cell parameters and cell volume of the activated materials contrasted with unblemished TiO_2 additionally inferred that neodymium particles were scattered on the outside of TiO_2 [287]. Also, the merging of Nd reformed the BET surface area of $Nd-TiO_2$ materials, contingent upon the quantity of Nd particles. Low loadings of Nd fundamentally amplified the TiO_2 surface region, which was an acceptable concurrence with the outcomes acquired [287]. Interestingly, a generously proportioned amount of Nd diminished the TiO_2 BET surface region, likely because of the blockage of a portion of the pores by the Nd clusters. Alteration of TiO_2 with neodymium altogether affected the material surface, prompting increments in the BET surface zone and micropore volume. It was additionally revealed that alteration of titania with neodymium enormously influenced the light absorption property of the photocatalysts [288]. The incorporation of Nd broadened the absorption range of TiO_2 into the visible region and diminished the titania band gap, which was allocated to the charge transfer between the Nd particle 4f level and the titania conduction or valence band [289]. A comparative perception was accounted for

TABLE 2.3

The Structural and Physical Properties of the Various Tm-TiO$_2$ Photocatalysts

Sr. No.	Preparation Method/Obtained Structure	Chemical Characteristics of RE Species	BET Surface Area (m²/g)	Band-gap (eV)	Unit Cell Parameter	Photocatalytic Activity Setup	Upconversion Measurement
1	Sol–gel Nanoparticles	Ion	–	3.25-pristine TiO$_2$ 2.81-ErTiO$_2$	–0.1360-pristine TiO$_2$ 0.1365-Ho-TiO$_2$	Model contaminant: methyl orange Irradiation source: Mercury lamp	Not tested
2	Sol–gel Nanoparticles	Ion	38.42-pristine TiO$_2$ 76.76 for-Ho-TiO$_2$	No data	0.1234-pristine TiO$_2$ 0.1245-Ho-TiO$_2$	Model contaminant: methyl orange Irradiation source and range: 500 W high-pressure mercury lamp	
3	Anodization Nanotubes	Oxide		3.35-pristine TiO$_2$ 3.30-ErTiO$_2$	136.64-pristine TiO$_2$ 136.73-Ho-TiO$_2$	Model contaminant: Phenol and toluene Irradiation source and range: 1000 W Xe lamp and LEDs (465 nm)	λ_{ex} = 375, 530, 650 and 980 nm λ_{em} = not observed
4	Sol–gel Nanoparticles	Ion			0.1225-pristine TiO$_2$ 0.1235-Ho-TiO$_2$	Model contaminant: methyl orange Irradiation source: 500 W high-pressure mercury lamp	
5	Hydrothermal Nanowires	Ion			Model contaminant: methylene blue Irradiation source and range: 175 W high-pressure mercury lamp	Model contaminant: methylene blue Irradiation source and range: 175 W high-pressure mercury lamp	

improved absorption properties of Nd-TiO$_2$ materials contrasted with perfect TiO$_2$ and experiential five typical absorption peaks situated at 527, 586, 762, 809 and 862 nm, which perhaps identified with the 4f electron transition or the f-f transition of the Nd particles [290].

Besides, it was indicated that neodymium activation can influence the morphology of the TiO$_2$ particles. Pure TiO$_2$ exhibited intermittently sized and formed grains, while on account of Nd-activated materials, grains with a progressively consistent size and shape were experiential. The pure titania nanoparticles produced accumulates, while the fuse of the neodymium dopant in the amounts of 1 at.% and 3 at.% generated the TiO$_2$ accumulates to decrease [284]. Adding together of neodymium ions has been revealed to enhance the photocatalytic activity of titania for the confiscation of methyl orange, phenol, Remazol black B, direct blue 53, reactive brilliant red X-3B, 2-mercaptobenzothiazole, malachite green and rhodamine B, in addition to the photo reduction of chromium (VI) as displayed in Table 2.4. It was accounted for that a proper content of Nd-dopant assumed an essential job in determining the photocatalytic activity of Nd-TiO$_2$ materials approaching methyl orange disintegration, and an ideal amount of dopant in this framework equaled to 3% neodymium. The utmost activity was witnessed for TiO$_2$ comprising 1% of Nd, which was 30% superior to that for the undoped TiO$_2$. It is likewise fascinating to take note of that Gomez et al. [280] investigated the fundamental components influencing the phenol debasement were measure of neodymium and the band-gap structure, while the surface area was the urgent boundary on account of rhodamine B corruption. Moreover, alteration of TiO$_2$ with neodymium led to defects in the lattice, which went about as electron traps, bringing about an improved detachment of charge carriers [54]. These outcomes are nearly similar with the outcomes observed by Nassoko et al. [291], who investigated that Nd incorporated into the TiO$_2$ lattice, and under visible light irradiation the electrons were excited from the valence band to the Nd^{3+}/Nd^{2+} doping energy level. Moreover, investigation established that hydroxyl radicals, classified among reactive species, were created by photo-generated electrons both from Nd-doped titania and rhodamine B and were mostly accountable for the photo degradation of the dye molecules [291]. In one more case, Parnicka et al. [283] performed estimations of photocatalytic debasement within the sight of scroungers, and the outcomes inferred that e$^-$ and O^{2-} were answerable for the photocatalytic corruption of phenol within the sight of Nd-TiO$_2$ tests under apparent light illumination. Moreover, the action spectra analysis showed that Nd-TiO$_2$ tests could be enacted by visible light with a frequency range from 400 to 480 nm, proposing that the upconversion cycle of process of visible-to-ultraviolet light not liable for the visible light-determined photoactivity of Nd-TiO$_2$ photocatalysts [292]. The preparation method, chemical characteristic, BET surface area, structural parameters, energy band gap, photocatalytic and photoluminescence properties of various Nd^{3+} doped TiO$_2$ materials was given in Table 2.4.

TABLE 2.4

The Preparation Method, Chemical Characteristic, BET Surface Area, Structural Parameters, Energy Band Gap, Photocatalytic and Photoluminescence Properties of Various Nd^{3+}-Doped TiO_2 Materials

Sl. No.	Preparation Method/Obtained Structure	Chemical Characteristics of RE Species	BET Surface Area (m²/g)	Unit Cell Parameters	Band Gap (eV)	Photocatalytic Activity	Up-Conversion Measurements
1.	Sol–gel/nanoparticles	Ion, oxide	108.7-pristine TiO_2 123.4- Nd-TiO_2	a = 3.786 nm, b = c = 9.507 nm, V = 136.315 Å³, pristine TiO_2 a = 3.778 nm, b = c = 9.532 nm, V = 136.629 Å³-Nd-TiO_2	No data	**Model contaminant:** methyl orange **Irradiation source and range:** direct sunlight (4% UV and 43% visible light)	λ_{ex} = 345nm λ_{em}: ~410, ~450, ~475 and ~520 nm
2	Mild Microwave assisted/nanoparticles	Ion	239-pristine TiO_2 258- Nd-TiO_2	a = 3.80 Å, c = 9.64 Å-pristine TiO_2 a = 3.80 Å, c =9.61Å-Nd-TiO_2	2.95-pristine TiO_2 2.89-Nd-TiO_2	**Model contaminant:** phenol **Irradiation source and range:** medium-pressure mercury lamp (λ_{max} = 365 nm).	Not tested
3	Sol–gel and precipitation/nanoparticles	Ion, oxide	81- pristine TiO_2 127.1- Nd-TiO_2	No data	No data	**Model contaminant:** remazol black B **Irradiation source and range:** high-pressure mercury lamp (λ < 400 nm)	Not tested

(Continued)

TABLE 2.4 *(Continued)*

The Preparation Method, Chemical Characteristic, BET Surface Area, Structural Parameters, Energy Band Gap, Photocatalytic and Photoluminescence Properties of Various Nd^{3+}-Doped TiO_2 Materials

Sl. No.	Preparation Method/Obtained Structure	Chemical Characteristics of RE Species	BET Surface Area (m^2/g)	Unit Cell Parameters	Band Gap (eV)	Photocatalytic Activity	Up-Conversion Measurements
4	Sol–gel/ nanoparticles	Ion, Oxidation state	223-pristine TiO_2 312- Nd-TiO_2	No data	3.27- pristine TiO_2 3.25- Nd-TiO_2	**Model contaminant:** direct blue 53 **Irradiation source and range:** xenon lamp ($\lambda > 320$ nm)	Not tested
5	Sol–gel/monolith	Ion	72.6-ristine TiO_2 74.9- Nd-TiO_2	No data	3.10- pristine TiO_2 2.96- Nd-TiO_2	**Model contaminant:** methyl orange **Irradiation source and range:** Xenon lamp ($\lambda > 365$ nm)	Not tested
6	Sol–gel	Ion	50.2-pristine TiO_2 78.5- Nd-TiO_2	No data	No data	**Model contaminant:** 2-mercaptobenzothiazole **Irradiation source and range:** medium-pressure mercury lamp ($\lambda = 365$ nm)	Not tested
7	Sol–gel/ nanoparticles	Ion	No data	No data	No data	**Model contaminant:** methyl orange **Irradiation source and range:** UV–Vis 20W Phillips lamps	Not tested
8	Sol–gel	Ion	43.29-pristine $TiO_{290.91}$- Nd-TiO_2	No data	No data	**Model contaminant:** 2-mercaptobenzothiazole **Irradiation source and range:** high-pressure sodium lamp (400–800 nm)	$\lambda_{ex} = 325$nm $\lambda_{em} = 504, 539, 596$ and 522 nm

2.4 The Origin of RE-TiO$_2$ Visible Light Photoactivity

2.4.1 Basic Principles of the Theoretical Study of Lanthanides-TiO$_2$ Photocatalysts

Quantum chemical calculations are economical, and quick strategies can give a portrayal of the electronic and optical properties of surface-modified TiO$_2$ mechanisms at the plan stage (preceding preparation) of new photocatalysts. The outcomes from these hypothetical estimations may give significant data to comprehension and foresee the impact of RE metals on the electronic and nuclear structure of RE-activated TiO$_2$ properties [292]. The more ambitious advances that may be challenged for achieving that goal are associated with density functional theory (DFT) methods [292]. DFT theory can conquer the purpose of accompanying examining electronic structures in solid-state physics; more significantly, it is a compelling methodology for anticipating the pattern of the energy gap variety in doped semiconductors. In spite of the experimental investigations of RE-TiO$_2$ (as displayed in Tables 2.1–2.4), there is a difficult issue in the description of the impacts of impurities on the TiO$_2$ photocatalyst execution. Along these lines, hypothetical groups have revealed various computations to anticipate the properties of the intricate RE-TiO$_2$ systems to give more important knowledge into the test readiness that could work on the photo response movement from the UV to the Vis light area (as summed up in Table 2.5). Based on the primary standards estimations, the hypothetical outcomes show that the confined 4f conditions of lanthanide particles (RE^{3+}) ordinarily bring about band gap narrowing and photo response upgrades under the visible light region [293]. Therefore, in this part, we summed up the hypothetical estimation of the electronic structure, charge density, and optical properties of RE^{3+}-activated TiO$_2$ systems.

The component of the improved band-gap narrowing and the related optical properties are likewise talked about in this part. It may be noticed very well that up to this second hypothetical examinations on RE=Nd^{3+}, Tm^{3+}, Ho^{3+} and Er^{3+} doping are relatively rare.

2.4.2 Computational Characterization of the Investigated RE-TiO$_2$ Structures

One of computationally examined RE-TiO$_2$ system (i.e., Nd$_2$Ti$_2$O$_7$) was described by Bruyer and Sayede in 2010 [294]. The investigators designated that the Nd$_2$Ti$_2$O$_7$ system could be applied in the water-parting response and they detailed the outcomes from DFT techniques on the underlying improvement for Nd-TiO$_2$ in different atomic arrangements (Table 2.5) [294]. The room temperature arrangement of the examined structure was monoclinic *P*21 with the lattice constants $a=13.02$ Å, $b=5.48$ Å, $c=7.62$ Å and $\beta=98.28°$. Bruyer and Sayede gave outcomes on the chemical bonding, structural phase

TABLE 2.5

Computational Details for the Computed RE-TiO$_2$ Structures Available in the Literature

Type of structure	The k-point meshes used for geometrical optimizations	The plane-wave energy cutoff	The structural energy convergence/the forces of relaxation	Methods	Software
Nd$_2$TiO$_2$O$_7$	4 × 4 × 4 4 × 4 × 2 4 × 4 × 1	500 eV	self-consistent convergence of the total energy/ 0.01eV/Å	DFTGGA+U (U = 7 eV) with PW	VASP
Nd^{3+}-TiO$_2$	Not Provided	Not provided	Not provided	DFT	Not provided
Tm$_x$Ti$_{1-x}$O$_2$	x 7 × 3 2 × 3 × 3 x 3 × 3 2 × 4 × 1	400 eV	5.0 × 10^{-7} eV/atom/0.01 eV/Å	DFT/GGA with PAW	CASTEP
Ho-TiO$_2$	8 × 8 × 11 3 × 7 × 1 2 × 2 × 1	400 eV	<1 meV/atom <0.02 eV Å	DFT/GGA with PAW	VASP

stability and ferroelectronic properties, in addition to the electronic structure of the Nd-TiO$_2$ system. The electronic structures of Nd$_2$Ti$_2$O$_7$ have demonstrated via the general gradient approximation method (GGA+U method). On the basis of GGA+U calculations and the density of states (DOS), the authors recommended a half-metallic nature for Nd$_2$Ti$_2$O$_7$ [294]. Nonetheless, the GGA+U results moved the Nd 4f states ahead of the EF, acquiescent to an insulating ground state. The localized Nd 4f levels at the top of the valence band (VB) and at a lower position in the conduction band (CB) worked as charge-trapping sites, showing that the Nd$_2$Ti$_2$O$_7$ structure had a condensed photocatalytic activity [294]. Li et al. [295] experiential band-gap lessening for the 0%, 1% and 1.5% Nd^{3+}-doped TiO$_2$ nanoparticles (NPs). To understand the band-gap (Eg) transformation, the researchers used the generalized gradient approximation (GGA) with the linearized augmented plane-wave (LAPW) technique. The Nd-doped anatase was represented by a structurally ameliorated NdTi$_7$O$_{16}$ super cell with Nd^{3+} in a substitutional site. Li et al. [295] specified that the band gap narrowing was mainly ascribed to the substitutional Nd^{3+} ions. Anchored in the acquired calculations, the researchers show that Nd^{3+} ions incorporated electron states into the band gap of TiO$_2$ (~ 2.26 eV), and these electron states predisposed the new lowest vacant molecular orbitals (LVMOs). The calculated band gap of pristine TiO$_2$ capable of 1.97 eV, with the O 2p orbital, states involvement at the top of the VB and the Ti 3d orbital assistance at the CB [295]. The calculated Eg of Ti$_7$NdO$_{16}$ that serves as 4 at.% Nd equivalent to 1.97 eV, and from the DOS examination, according to the conclusions of the researchers several electronic states, which were positioned close to the CB, were introduced into the band gap of TiO$_2$ by Nd *4f* electrons to form the new LVMO. Consequently, the absorption edge transition for the doped material can be from O *2p* to Nd *4f* instead of Ti *3d*, as in pristine TiO$_2$. Li et al. found that the utmost band-gap reduction (Eg=0.55 eV) was found for 1.5 at.% Nd-doped TiO$_2$ nanoparticles. On the basis of acquired outcomes, the investigators indicated that Eg narrowing could be related to the substitutional Nd^{3+} ions, which incorporates electron states into the band gap of TiO$_2$ to construct the new LVMO [295].

Mazierski et al. [238] investigated the electronic structure and partial density of states (PDOS) of pristine TiO$_2$ and Ho-TiO$_2$. The researchers concluded the system consolidation subsequent to the configuration of defects at the surface of the Ho-anatase TiO$_2$(101) [238]. Wei and Jia [293] announced the electronic structure, charge density and optical properties of Tm^{3+}-activated TiO$_2$ dependent on DFT calculations (Table 2.5). The researchers showed three unique arrangements of the Ti$_{1-x}$Tm$_x$O$_2$ systems, in which the variety x was characterized by the upsides of 0.0417, 0.0625, and 0.125. Tm molecules in Ti$_{1-x}$Tm$_x$O$_2$ systems were brought into the standard Ti lattice position. The supercells were made out of products of the lattice vectors a, b, c, for example, 3×2×1 (72 atoms), 2×2×1 (48 atoms) and 2×1×1 (24 atoms) supercells for the anatase phase Ti$_{1-x}$Tm$_x$O$_2$ system, which symbolized the doping concentrations of 1.39 at.% (Ti$_{23}$O$_{48}$Tm$_1$), 2.08 at.% (Ti$_{15}$O$_{32}$Tm$_1$) and 4.17 at.% (Ti$_7$O$_{16}$Tm$_1$), respectively. The optical properties and electronic structures of the Tm-TiO2 systems were examined utilizing the plane-wave DFT technique with the Padrew-Burke-Ernzerh of generalized gradient approximation functional (PBE/GGA) [293]. Wei and Jia [293] talked about the component of the improved band-gap (Eg) narrowing and the related optical properties of the researched Tm-TiO$_2$ system. Wei and Jia noticed the decrease of the band gap that contrasted with unadulterated anatase TiO$_2$ because of the introduction of Tm into the anatase TiO$_2$ structure. The band gap was evidently diminished from 2.17 to 1.93 eV in the Tm-activated Ti$_{23}$O$_{48}$Tm$_1$ system. Therefore, Tm doping limited the band gap to a worth of roughly ΔEg=0.24 eV [293]. For the Tm-activated anatase Ti$_{15}$O$_{32}$Tm$_1$ and Ti$_7$O$_{16}$Tm$_1$, the band-gap energies, that is, 1.92 eV (ΔEg=0.22) and 1.83 eV (ΔEg=0.29), compared to the doping levels of 2.08 at.% and 4.17 at.% of Tm particles, separately. Besides, the introduced outcomes demonstrated that the Tm iotas considerably modified the DOS, inciting varieties in both the VB and the CB [293]. The researchers concluded that the substitutional Tm dopant expanded the delocalized Tm 4f states over the VB, which limits the band gap, bringing about a red shift of the optical absorption edges [293]. A reliance between the Tm concentration and the band-gap decrease was noticed. The introduced study was the beginning stage for the planned band-gap decrease utilizing Tm-activated TiO$_2$ for broadening the optical absorption into the visible solar light region for improved photocatalytic performance [293]. From the considered models talked about in this section, apparently the hypothetical description at the DFT level of the RE-TiO$_2$ semiconductor oxides presents various issues.

Acquiring more exact deformity arrangement energies and a more reliable distribution of the excess electrons requires going beyond the GGA in DFT by using hybrid approaches; in general, this implies expanding the computational requests. Consequently, the decision of the model of the diminished system may become crucial. The question arises whether generally utilized intermittent models will be helpful in this regard. Ongoing examinations on the electronic structure of defective TiO_2 (110) surfaces utilizing intermittent models and hybrid functionals in [296] show that an expanded interest can be anticipated in this kind of estimation in the near future. Be that as it may, cluster approaches will in any case be valuable (and less expensive computationally).

Regardless, a general hybrid functional that incorporates a fixed measure of the specific trade may not really have the option to describe the electronic structure of all $RE-TiO_2$ structures with a similar precision.

2.4.3 Anti-Stokes Upconversion Process in the RE-TiO$_2$ Mechanism

To clarify the conceivable system of lanthanide-activated TiO_2 excitation under visible light, the photo degradation of model toxins ought to be researched as a component of irradiation wavelength (activity spectra (AS) examination) [261]. In view of AS examination, it is feasible to figure out which part of light consumed by the photocatalyst is associated with the photocatalytic responses [286]. If the upconversion measure is associated with the improvement of the photocatalytic movement under visible light, then at that point an upgrade of the photocatalytic action in RE absorption bands locale ought to be noticed. In the recent literature, there are a couple of reports of $RE-TiO_2$ AS estimations. Parnicka et al. [286] examined the impact of the light frequency on the photocatalytic movement of the phenol corruption response within the sight of $Nd-TiO_2$ photocatalysts. The AS of the examples didn't take after the $Nd-TiO_2$ absorption spectra. The Nd-activated TiO_2 photocatalyst showed photocatalytic action under visible light in a consistent irradiation range from 400 to 480 nm [286]. Photocatalytic activity under the Nd absorption band wavelengths ($\lambda = 525, 585, 745$ and 805 nm) has not been noted.

No relationship between the frequencies liable for the photocatalytic activity and the absorption spectra of RE was likewise seen by Reszczyńska et al. [265]; the photocatalysts $Er-Yb-TiO_2$, $Nd-Eu-TiO_2$ and $Eu-Ho-TiO_2$ showed photoactivity in 420–475 nm, 420–450 nm and 420–450 nm regions, respectively. The improvement in the photocatalytic activity under 488, 522, 524, 586 or 653 nm was not noticed [265]. As indicated by the ongoing literature, the explanation that the effect of the upconversion process on the photocatalytic activity was not generally noticed was a result of the enormous distance between the ground and excited states of the RE particles (multiphoton excitation) and multiphonon relaxation processes that were influenced by the properties of the crystal structure of the photocatalysts [297]. Also, the aqueous nature of the solutions caused the association of RE with water (both in the internal and external coordination spheres of the RE[3+]) and prompted a serious extinguishing of the RE iridescence by means of O–H vibrations [298]. The most ideal approach to limit the water deactivation procedures is to shield the RE[3+] particle from solvent communications, for example, with a chelating cycle or by restricting RE with peptides [298]. In the recent literature, there is no data in regard to TiO_2 photocatalysts activated by RE particles with additional assurance from aqueous interactions. In view of the recent study, there is no dependable verification that the upconversion procedure truly affects the RE-TiO_2 photocatalytic activity under visible light irradiation, hence this viewpoint ought to be explained later. Wang et al. [285] expressed that the fuse of Ho or Nd particles into the TiO_2 structure prompts the development of sub-band-gap states that lie beneath the CB of the semiconductor. The electrons can be excited from the VB to the lanthanide 4f level under visible light irradiation. These trapped electrons can respond with O_2 and structure receptive oxygen species, like O_2, HO_2 and H_2O_2, prompting organic pollutant degradation [238].

2.5 Conclusion and Perspectives

This chapter gives an idea about comprehensive summary of the examinations that explored conversion of TiO_2 with Er, Ho, Nd and Tm particles. Lanthanide particles were utilized for the TiO_2 change to acquire photocatalysts being dynamic under visible light irradiation. Specifically, dye degradation has been discovered to be a well-known decision for the correlation of the photoactivity between pristine TiO_2 and RE-TiO_2. Information of the previously mentioned articles recommends that RE-TiO_2 conversion is compelling in expanding the photocatalytic activity of TiO_2, particularly at low loading levels. Nonetheless, due to different photocatalytic activity setups picked for the conveyed tests, the decision of a lanthanide causing the most noteworthy increment is testing. As indicated by the previously published articles, there is no dependable verification that the upconversion procedure truly affects the RE-TiO_2 photocatalytic movement under a visible light irradiation. Also, there is presently no data in regard to TiO_2 photocatalysts tailored with RE particles with additional insurance from aqueous interactions in the latest published articles. Accordingly, those two perspectives ought to be explained later. Likewise, time-consuming and often very expensive experimental testing, just as considerable amount of the conceivable RE-TiO_2 structure combinations, leads to limitations in proper characterization of RE-TiO_2 photocatalysts. It has been broadly demonstrated that hypothetical computations joined with chemo informatics techniques (for example, Quantitative Structure-Property Relationship displaying, QSPR) may uphold the trial testing of new potential photocatalysts. In any case, a powerful methodology that consolidates hypothetical and computational estimations with trial configuration has not much grown very much at this point. Since getting sorted out a synergistic collaboration among experimentalists and theoreticians is a basic advance, we proposed a guide showing which targets should be accomplished in a short-, medium-, and long-haul undertakings. In the initial step (for example, momentary undertaking) the accompanying errands is of most elevated significance: i) determination of characterized materials for the analyses; ii) choice and estimations of movement important endpoints. While analyzing photocatalytic properties of RE-TiO$_2$ systems, it is significant that electronic and nuclear properties of TiO$_2$, RE particles and RE-TiO$_2$ systems are characterized materials and seen, so legitimate derivations can be acquired, and how these structure properties impact the photocatalytic activity. Hence, in the subsequent advancement, a computational methodology ought to be embraced, where quite possibly the main parts of hypothetical techniques are created. The physicochemical highlights, characterized as descriptors, can help in understanding the connections between nuclear structure and electronic/optical properties of researched subatomic models. In the drawn out task, the information base contained physico-substance properties and photocatalytic action estimations of examined TiO$_2$ systems ought to be created. A while later, on a premise of the assembled information, computational models (for example, QSPR models for nanoparticles, nano-QSPR) for dependable endpoints ought to be created, giving an information how the photocatalyst structure influences its ideal property (for example, photocatalytic action). At last, the created models ought to be utilized to begin with appraise photocatalytic movement of new (untested) structures at beginning phase of trial plan. The drawn out objective is to make nano-QSPR models of TiO$_2$-based structures to produce adequate exploratory, and robotic information for test scientists that will help choosing which underlying highlights ought to be utilized as well as changed to plan novel and proficient photocatalytic systems having desired properties.

.

3

Synthesis of Titanate Materials

From the past decades, titanate-based perovskite materials have received much attention among research community due to their exciting properties, including ferroelectric, dielectric, ionic conduction, optoelectronic, superconductivity, etc. [299–304]. Hence, tremendous efforts devoted to synthesizing the titanate materials are phenomenal owing to their applications in various significant fields, such as sensing, photo-thermal therapy, bio-imaging, catalysis, solar cells and so on [305–308]. To date, much progress has been made to synthesize perovskite materials with desired sizes, compositions, facets/shapes and architectures. The synthesizing techniques preferably comprise the following characteristics: (i) fabrication of material with appropriate composition and characteristics, (ii) resource and energy efficient, (iii) rapid, flexible, continuous and use preferably one step (iv) large-scale production and (v) minimal hazards to workers or end users. Many synthesis techniques for the fabrication of titanate-based perovskite materials, such as hydrothermal, coprecipitation, sol–gel, sonochemical, solution combustion, solid-state reaction, pulsed laser method and sputtering method were addressed by meeting all the above requirements for safe syntheses. Schematic to show different synthesis routes followed to prepare titanium-centered perovskites is depicted in Figure 3.1. This review provides a comprehensive insight into the various aspects of methodology for synthesis of titanate-centered perovskite materials.

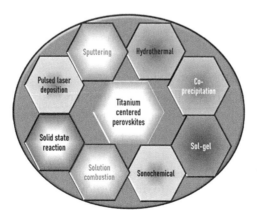

FIGURE 3.1
Different synthesis routes followed to prepare titanium-centered perovskites.

DOI: 10.1201/9781003381907-3   47

3.1 Hydrothermal Method

In the early 19th century, a British Geologist Sir Roderick Murchison first demonstrated the hydrothermal method to explain the formation of variety of minerals and rocks due to elevated temperature and pressure of water in the Earth's crust. Based on this idea, artificial hydrothermal method was developed for the synthesis of various nanostructured materials by dissolution and crystallization process. Hydrothermal route is a wet-chemistry technique involving heterogeneous chemical reactions carried out in a sealed hydrothermal vessels called autoclave. This route benefits from high pressure developed by heating precursor solutions in a sealed autoclave by maintaining high temperature than boiling point of the solvents results in chemical transport reaction that depends on transport of reactants from liquid phase to nucleation of desired product. In a typical method, temperature plays a significant role in reaction kinetics for the formation of desired products and achieve thermodynamically stable product phase [309–313]. However, pressure is also a necessary requisite for the solubility, limiting the super saturation results in crystallization process and quality product.

Whereas, time is also a requisite factor since the short and long duration syntheses processes favored in kinetically and thermodynamically stable phases, respectively, which was equivalent to chosen pressure-temperature regime. In addition, various experimental parameters of the synthesis, namely composition and concentration of the starting materials, stabilizing agents, solvents, pH of the reaction solution and geometry of autoclaves, offer important effects on the products [313]. Figure 3.2 shows the experimental parameters monitored to synthesis titanium-centered ABO_3 perovskites. However, hydrothermal route presents significant advantages, including

- Relatively trivial synthesis conditions
- Ecofriendly
- Single-step synthesis procedure
- Economical in terms of the instrumentation
- A good control over composition of the final products
- Less contamination of the product with surrounding environment.

The synthesis of titanate-centered perovskites gains much interest for the research community due to their unique properties, which mainly depends on morphological features, namely grain size, porosity, etc. From

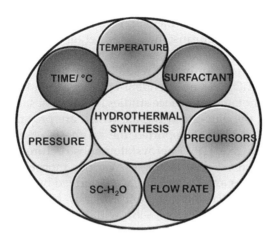

FIGURE 3.2
Experimental parameters monitored to synthesis titanium-centered ABO_3 perovskites.

the past decades, a wide range of titanate perovskites have been fabricated via hydrothermal route and discussed. Lee et al. [314] synthesized $BaTiO_3$ nanocrystalline powders by hydrothermal method using titanium acylate and barium acetate. Titanium acylate was prepared by dissolve titanium (IV) isopropoxide with glacial acetic acid at $\sim 25°C$ for 1 h. Barium acetate (0.21 mol) was dissolved in ~ 100 ml of deionized water and added slowly to the obtained titanium acylate solution under N_2 atmosphere. The pH of the solution was controlled by varying KOH. All reaction precursors were taken in an autoclave (250 ml) maintained at $\sim 120°C$. The obtained product was cleaned and dried at $100°C$ for 12 h in an oven. It was found that pH of the solution controls the size, crystallinity and morphology of $BaTiO_3$ powders. Typically, Qi Feng et al. [315] reported the synthesis of plate-like-structured $BaTiO_3$ and anatase with high degree of crystal-axis orientation via hydrothermal route. A layered titanate of $K_{0.8}Ti_{1.73}Li_{0.27}O_4$ with a lepidocrocite-like layered structure, which has a platelike particle morphology, was used as a precursor. The lepidocrocite-like layered $K_{0.8}Ti_{1.73}Li_{0.27}O_4$ was mixed with 1 M HNO_3 solution (1 L) for ~ 24 h to exchange K^+ and Li^+ in the layered structure with H^+. Later, layered titanate with H^+-form was dissolved with appropriated quantity of $Ba(OH)_2$ solution (0, 0.1, 0.2 and 0.3 M) and distilled water. All the precursors were placed in a Teflon-lined, sealed stainless steel vessel (30 mL) and then hydrothermally treated at 150 or 200°C for 24 h. The obtained product was filtered, washed with hot distilled water and dried at 80°C. Formation mechanism involved in the layered plate-like-structured $BaTiO_3$ was an in situ topotactic transformation reaction and dissolution-deposition on particle surface. The more prominent topotactic transformation reaction at low $Ba(OH)_2$ concentration was noticed during the formation of $BaTiO_3$. For instance, Souza et al. [316] synthesized $SrTiO_3$ nanoparticles by using microwave-assisted hydrothermal route by varying reaction time from 4 to 160 min. The results showed that crystalline phase of perovskite-type $SrTiO_3$ was independent of the reaction time. The detailed investigation on photoluminescence studies reveals that samples emit maximum intensity in the blue-green region, which are prepared with shorter reaction time. Lu et al. [317] reported the synthesis of lead-free bismuth sodium titanate ($Bi_{0.5}Na_{0.5}TiO_3$) perovskite nanostructures by a hydrothermal route without any use of templates. Various morphologies, including nanowires, nanocubes and nanoplates, were derived by controlled reaction time duration and NaOH concentration. Typically, nanoplates with 400 nm length were achieved for 200°C for ~48 h with NaOH concentration of 8–12 M. In addition, uniformed cubes and nanowires were tuned by varying NaOH concentration and reaction time (~60 h).

The systematic morphological growth mechanisms of the $Bi_{0.5}Na_{0.5}TiO_3$ nanostructures were proposed. The formation mechanism was explained in two steps; (i) transformation of amorphous precursors into polycrystalline particles via water elimination and (ii) dissolution and recrystallization-dissolution of the precursors forms aqueous metal species followed by recrystallization from super saturated solution. Further, the growth behavior of $Bi_{0.5}Na_{0.5}TiO_3$ nanowires via oriented attachment mechanism as well as crystal splitting was proposed. Cao and coworkers [318] have prepared barium titanate ($BaTiO_3$) nano-maces via hydrothermal route by control the temperature and alkalinity. The detailed study of morphological transformation mechanism confessed, which mainly involves the generation of chemical site, prompting the ion exchange reaction and the dissolution–precipitation reaction (Figure 3.3). Finally, the authors reveal that the reaction time cannot offer enough driving force to endorse the formation of tetragonal $BaTiO_3$ nanostructures.

Wang et al. [319] synthesized a gradient $BaTiO_3$-$Ba_{1-x}Sr_xTiO_3$ core–shell nanoparticles by employing hydrothermal approach using $Sr(OH)_2.8H_2O$, HCl and ammonium hydroxide solution titanium, butoxide and ethanol as starting materials. The variations in the intensities and peak position of XRD confirms the formation of the shell layer, which depends on hydrothermal reaction duration. SEM images of as-received $BaTiO_3$ nanoparticles reveals cubic or spherical nanoparticles with size of 110 nm. However, as-prepared nanoparticles also exhibit similar morphology and size even after Ba dissolution. This specifies the formation of Ba-deficient surface layer. Subsequently, the size and morphology of core–shell structure did not exhibit any noticeable alterations irrespective of reaction time noticed (Figure 3.4a). From TEM and EDS, it was found that the surface modification of nanoparticle by increasing the Sr quantity is more efficient than that at the center (Figure 3.4b).

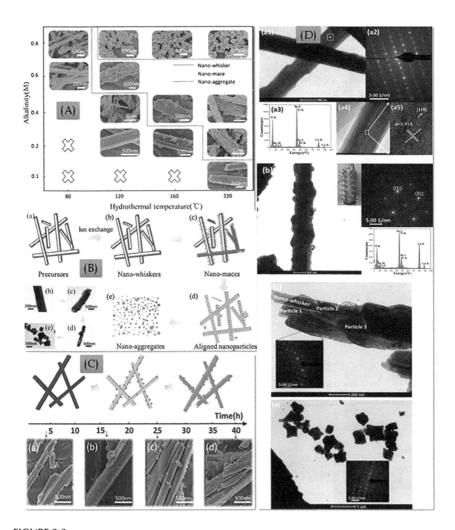

FIGURE 3.3
(a) Morphology evolution diagram of BaTiO₃ nanostructures with various as-temperature and alkalinity for fixed time (16 h). (b) Proposed schematic of morphological evolution and concentration of alkaline (0.2–0.8 M)-dependent morphology of BaTiO₃ nanostructures. (c) BaTiO₃ nanostructures synthesized by varying reaction time (4, 16, 25 and 40 h) with fixed 160°C/0.2 M, (d) TEM, HRTEM images, SAED and EDS of BaTiO₃ nano-whiskers. (Reprinted with the permission from Ref. [318]. Copyright © Elsevier publications.)

The shape-controlled SrTiO₃ crystallites were synthesized by Xu et al. [320] via sol–gel-hydrothermal route. SEM and TEM observations clearly evidence shape evolutions at prolonged reaction time by unfeatured conglomeration through eight-pod star-shaped particles to cubic crystals

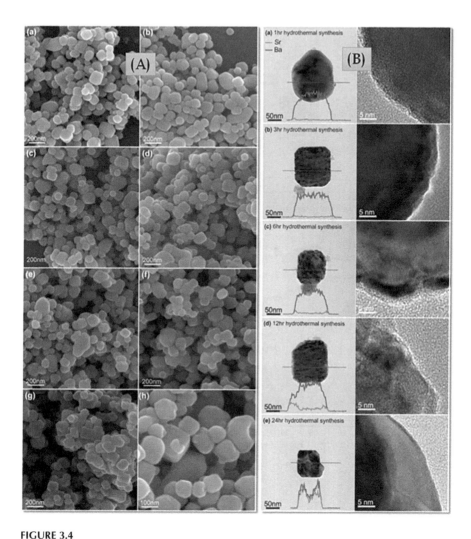

FIGURE 3.4
(a) SEM images of as-received, as-prepared, core–shell BaTiO₃ nanoparticles prepared with different reaction time (1, 3, 6, 12 and 24 h) at 200°C and (b) TEM images and EDS line mapping of the core–shell nanoparticles. (Reprinted with the permission from Ref. [319]. Copyright © Elsevier publications.)

(Figure 3.5a–c). The growth mechanism of the SrTiO₃ involves formation of nuclei, kinetically controlled reaction and orientations along [1 1 1] direction (Figure 3.5d).

Details of various works on titanium-based perovskites synthesized via hydrothermal method are presented in Table 3.1.

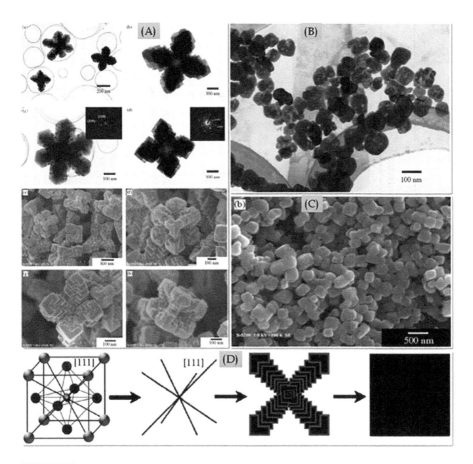

FIGURE 3.5
(a) TEM and SEM images of the product synthesized at 200°C for 12 h, (b, c) TEM and SEM images of sample prepared at 200°C for 12 h and 200°C for 24 h without acetic acid, respectively, and (d) Schematic illustration of the shape evolution of SrTiO₃ crystallites. (Reprinted with the permission from Ref. 320. Copyright © Elsevier publications.)

3.2 Coprecipitation

In 1981, Massart reported preparation of metal nanoparticles by facile and convenient coprecipitation technique. The coprecipitation route is a wet chemical method, which is an excellent choice for preparing nanoparticles from a supersaturated solution by tuning the pH level when greater purity and improved stoichiometric control are necessity [335]. The synthesis of

TABLE 3.1

List of Titanium-Centered ABO_3 Structured Perovskites Synthesized by Hydrothermal Route

Sl. No.	Materials	Precursors Used	Temperature (°C)	Phase	Size and Morphology	Characterizations	References
1	$SrTiO_3$	P25 TiO_2, NaOH $Sr(OH)_2 \cdot 8H_2O$	220	Cubic	Polygonal particles	PXRD, FESEM, UV-visible	Gao et al. [321]
2	$SrTiO_3$: Er^{3+}, Yb^{3+}	$Sr(NO_3)_2$, KOH $Er(NO_3)_3 \cdot 5H_2O$, P25 TiO_2	180	Cubic	20–50 nm; Nanocubes	PXRD, SEM, TEM, HRTEM, EDS, UC-PL	Xiao et al. [322]
3	$SrTiO_3$	$TiCl_4$, NaOH, $Sr(OH)_2 \cdot 6H_2O$ Lithiated sarcosine	170	Cubic	50 nm; Nanocubic	XRD, FESEM, TEM, SAED, ^{13}C NMR	Sreedhar et al. [323]
4	$CaTiO_3$	$CaCl_2$, NaOH, Titanium n-butoxide	180	Orthorhombic	53–130 nm; Prismatic particles	XRD, DTA-TG, DRS, FTIR, SEM, TEM, HRTEM	Stoyanova et al. [324]
5	$CaTiO_3$:Pr^{3+}	Ti $(OC_3H_7)_4$, $CaCl_2.2H_2O$, Pr $(NO_3)_3 \cdot 6H_2O$, KOH	140	Orthorhombic	Micro-sized cubes	PXRD, FT-Raman, FESEM, PL	Gonçalves et al. [325]
6	$SrTiO_3$:Fe	$Sr(CH_3CO_2)_2$, $Fe(NO_3)_3 \cdot 9H_2O$, TiO_2, KOH	800	Cubic	33–48 nm; Cubic-like shape	PXRD, SEM, TEM, XPS, VSM	Karaphun et al. [326]
7	$CaTiO_3$:Sm^{3+}	$Ti(OC_3H_7)_4$, $CaCl_2.2H_2O$, Sm_2O_3, KOH	140	Orthorhombic	Micro-cube	XRD, Micro-Raman scattering (MRS), DRS, PL, FESEM	Pinatti et al. [327]
8	$CaTiO_3$:Eu^{3+}	$TiCl_4$, $CaCl_2.2H_2O$, Eu_2O_3, KOH	140	Orthorhombic	Micro-cube like cages	PXRD, Raman, DRS, PL, FESEM	Mazzo et al. [328]

(Continued)

TABLE 3.1 (*Continued*)

List of Titanium-Centered ABO$_3$ Structured Perovskites Synthesized by Hydrothermal Route

Sl. No.	Materials	Precursors Used	Temperature (°C)	Phase	Size and Morphology	Characterizations	References
9	BaTiO$_3$	Ba(OH)$_2$·8H$_2$O, Tetrabutyl titanate, CTAB, KF, KCl, KBr, KI	180	Cubic	Coral-like architecture	XRD, FESEM, EDS, UV-Vis, PL	Zhao et al. [329]
10	BaTiO$_3$:Fe	Barium acetate, NaOH tetra-n-butyl orthotitanate, Ferric chloride	180	Cubic/ tetragonal	Nano-rod	XRD, TEM, PL, Raman, VSM	Verma et al. [330]
11	BaTiO$_3$:Eu^{3+} / Eu^{2+}	Ba(CH$_3$COO)$_2$, Eu(NO$_3$)$_3$, Ti(OC$_4$H$_9$)$_4$, ethylene glycol	180	Tetragonal	Dumbbell-like shape	XRD, SEM, TEM, PL, Raman	Feng et al. [331]
12	SrTiO$_3$	Sr(CH$_3$COO)$_2$, Ti[OCH(CH$_3$)$_2$]$_4$, EtOH, Citric acid, poly ethylene glycol	120	Cubic	Sponge-like structure	XRD, FESEM, EDX, FT-IR, Raman scattering, PL, UV-Vis, Electrochemical workstation	Jayabal et al. [332]
13	SrTiO$_3$/TiO$_2$	Ti(SO$_4$)$_2$, Polyvinyl pyrrolidone, NaOH, Sr(NO$_3$)$_2$	700	-	Microspheres	SEM, HRTEM, SAED, XRD, UV-Vis, Photo voltage spectroscopy	Wang et al. [333]
14	BaTiO$_3$	Ba(OH)$_2$ 0.8H$_2$O, TiO$_2$	100	Cubic	31–50 nm Spherical, hexagonal, rods	UV-Vis, PL, FT-IR, XRD, TEM, SEM	Patil et al. [334]
15	BaTiO$_3$	BaCl$_2$.2H$_2$O, Cl$_6$H$_{36}$O$_4$Ti, Carbon fiber, ZnCl$_2$	700	-	Homogeneous particle	XRD, SEM, EDX, DRS, PL, FT-IR	Demircivi et al. [303]

materials via coprecipitation reactions involve nucleation, slow growth, coarsening and/or agglomeration processes [336,337]. Coprecipitation reactions show the following characteristics:

- The obtained materials are normally insoluble produced from super-saturated solution.
- A large number of small building blocks were formed through nucleation process.
- The aggregation and Ostwald ripening process greatly affects the materials in terms of morphology and properties.
- The supersaturation conditions can be achieved on the basis of a chemical reaction.

$$XAy^+_{(aq)} + yBx^-_{(aq)} \rightarrow AxBy_{(s)}$$

Coprecipitation route deals with major benefits, namely easy and fast reaction, size and compositions of the materials can be easily controllable, homogeneous, less temperature, fewer energy consumption and no use of organic solvent. However, present method deals with some limitations, including required trained person for maintenance and regeneration, time consuming, generation of toxic liquid waste, problem of industrial production, necessity of alteration of pH and chance of precipitation of impurities with the product [338,339].

Xu et al. [340] reported the preparation of g-C_3N_4-coated $SrTiO_3$ as an active photocatalyst by coprecipitation route simply by using strontium nitrate, titanium isopropoxide, ethanol, nitric acid and ammonium solution as starting reagents. The typical procedure involves stoichiometric quantity of strontium nitrate, and titanium isopropoxide were well dissolved in ethanol solution. The 1 M nitric acid was slowly added to the reaction precursors to avoid the precipitation. The resultant solution was dissolved slowly in an ammonium solution (1 M) under continuous agitation. Finally, precipitation obtained at the end of the reaction was dried at 200°C and calcined at 800°C. Active $SrTiO_3$-C_3N_4 composites were synthesized by appropriate amount of $SrTiO_3$ and urea were finely ground using a mortar and pestle (~1 h). The obtained mixtures calcined at 400°C and 600°C in an alumina crucible. The characteristic PXRD peak at $2\theta=27.4°$ of g-C_3N_4 was noticed in $SrTiO_3$-C_3N_4 composite, demonstrating the formation of g-C_3N_4 in the $SrTiO_3$ during urea decomposition. Similarly, Tiwari et al. [341] synthesized Er^{3+}/Yb^{3+}-doped ZnO–$CaTiO_3$ nano-composite phosphor via coprecipitation method exhibiting up-conversion emission and temperature-sensing property. The composite phase was confirmed by XRD and EDAX results when above 10 mw % of ZnO doping. The characteristic emission peaks at ~410, 492, 524, 545 and 662 nm under 980 nm excitation were noticed. The 30 mw % of ZnO-doped samples show highest emission intensity. They found the green emission optimized sample with high color purity. The obtained results clearly evidence that the composite phosphor may be used as display devices and LEDs. Zhang and coworkers [342] reported $CaTiO_3:Pr^{3+}$ red-emitting nanophosphors via coprecipitation route. The luminescence emission of the prepared nanophosphors exhibits enhanced phosphorescence for the sample prepared at 700°C, which might be due to decline of trap numbers and increased recombination efficiency. The obtained results reveal the dependence of the fluorescence and phosphorescence on calcination temperature. Kumari et al. [343] prepared codoped $BaTiO_3:Eu^{3+}$, Yb^{3+} phosphor via coprecipitation route. PXRD profiles confirm the tetragonal $BaTiO_3$ phase. The up-conversion PL emission peaks in the range of 592–796 nm under ~980 nm diode laser excitation which was noticed. The band at 489 was assigned to transitions of Yb^{3+} ions. The broad band at 505 nm was noticed which may have attributed to presence of defect states in the products. They found that the CIE color coordinates tuned from green to orange color region display its applications in color tunable display devices. Mahata et al. [344] synthesized Er^{3+}/Yb^{3+} codoped $BaTiO_3$ phosphor by employing coprecipitation method. The good crystallinity with cubic phase of $BaTiO_3$ having size of 21–28 nm was estimated from PXRD results. The intensive green up-conversion emission was noticed by exciting the samples by 980 nm diode laser. The indirect band gap of the prepared sample was found to be ~2.9 eV. The color tuning from yellowish to green region with increase of excitation power was noticed.

3.3 Sol–gel

The sol–gel route was one of the well-developed synthesis approaches to prepare ceramic and glass materials in a wide range of forms, including ceramic fibers, ultrafine powders, microporous inorganic membranes, thin films, monolithic, etc. [345–347]. This typical preparation route involves few steps to obtain final products, such as hydrolysis, condensation and drying process. The formation mechanism undergoes to prepare nanomaterials via sol–gel method, including (i) rapid hydrolysis of initial metal salts, followed by sudden polycondensation to form three-dimensional gels by slow heating techniques, (ii) the process of polycondensation reactions proceeds during syneresis of gels to result a solid mass convoyed by contraction of gel network and removal of solvent from gel pores and (iii) simultaneously Ostwald ripening and phase transformations may also occur with syneresis. Later, the obtained gel is exposed to drying and heat-treatment process resulting in Xerogel or Aerogel.

The sol–gel route offers several advantages, namely control textural, surface and composition of materials in a molecular level, require low temperature, potential to prepare complex composition materials, produce high porosity and purity and homogeneity of the synthesized products [348,349]. However, this technique has its own limitations for the preparation of nanomaterials, including (i) upon drying, the wet gel gets contracted which results to fracture attributed to the production of great capillary stresses, (ii) avoid of residual porosity and OH groups was challenging and (iii) chances of preferential precipitation during the formation of sol due to various reactivity of alkoxide precursors. Cernea et al. [350] prepared cerium (5.5 mol%)-doped $BaTiO_3$ by a sol–gel route using barium acetate, cerium (III) acetylacetonate and titanium (IV) isopropoxide as initial materials. The nano-regime grains with size of ~140 nm exhibit an excellent affinity to agglomeration, which was clearly noticed in microstructure of dried gel. The decomposition of precursors through oxides and barium carbonate around 500°C was analyzed by many techniques. The presence of aggregates at higher temperature thanks to excellent bonding of Ti-OH. The amorphous nature of the sample at lower temperature, while highly crystalline tetragonal $BaTiO_3$ perovskite structure at 1,100°C, was observed. Recently, Hao et al. [351] reported three-dimensional (3D) $SrTiO_3$ hierarchical structure employed by TEG-sol technique under controlled atmospheric pressure. The well crystallized hollow-sphere architectures with sub-10 nm cubes was clearly evidenced. The alkalinity of the reaction solvent that plays a pivotal role in the formation of hierarchical architecture (formation and self-assembly of the nanocubes) was noticed (Figure 3.6a).

From SEM and TEM images, self-assembly of microspheres (0.2 to 1.1 μm) to form hierarchical $BaTiO_3$ microstructure was clearly noticed (Figure 3.6b, c). The prepared nanocubes show significant enhanced PL emission intensity due to large specific surface area (Figure 3.6d). The obtained results clearly reveal that the present method can offer mild reaction condition, quick formation and great efficiency to prepare high purity and industrial production of perovskite oxides with controlled morphology and fine nanostructures. Sun et al. [352] reported the core–shell SiO_2–$CaTiO_3$:Eu^{3+} structured phosphors fabricated by sol–gel process by coated layers of $CaTiO_3$:Eu^{3+} on spherical, monodisperse and nonaggregated SiO_2 particles with citric acid as a chelating agent. The PXRD, SEM and TEM clearly revealed layers of $CaTiO_3$:Eu^{3+} on SiO_2 particles (Figure 3.7a). The formation mechanism of core–shell structures including heterogeneous nucleation by homogeneous mixing of silica spheres into sol–gel precursors was clearly demonstrated in Figure 3.7b. The enhanced red color emission with increase of the shell thickness on the SiO_2 cores was noticed (Figure 3.7c). They clearly demonstrated that the following method was versatile to prepare variety of other homogeneous core–shell phosphors.

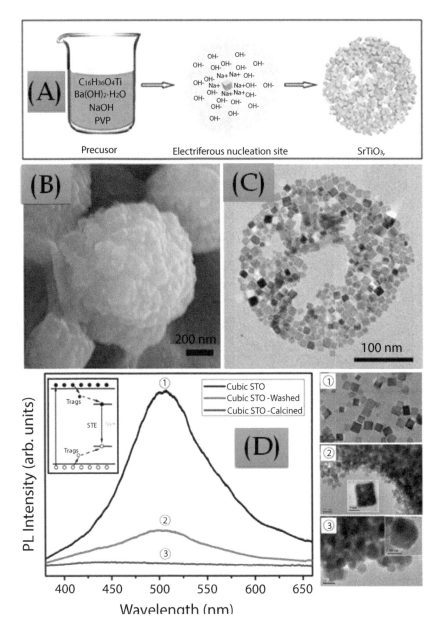

FIGURE 3.6
(a) Probable formation mechanism, (b, c) SEM and TEM images and (d) PL emission spectra and the corresponding structure illustration of the BaTiO$_3$ nanoparticles that synthesized at different alkaline conditions. (Reprinted with the permission from Ref. [351]. Copyright © Elsevier publications.)

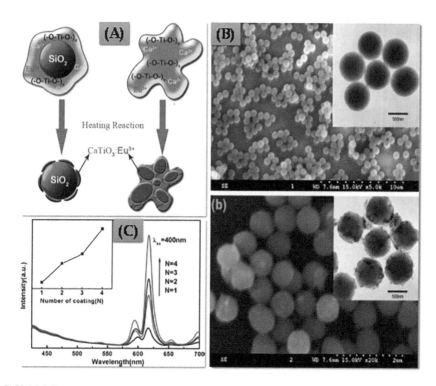

FIGURE 3.7

(a) Probable formation mechanism core–shell SiO_2–$CaTiO_3$: Eu^{3+} structured phosphors, (b) SEM micrographs of SiO_2 and core–shell. (Inset: corresponding TEM images.), and (c) PL emission spectra of core shell as a function of the number of coatings. (Reprinted with the permission from Ref. [352] Copyright © Elsevier publications.)

Many other reports are available on titanate-centered perovskites materials synthesized via sol–gel route, including $SrTiO_3$ [353], $CaTiO_3$:Yb, Er [354], $SrTiO_3$:Pr^{3+} [355], $CaTiO_3$: Nd^{3+} [356], $ATiO_3$ (A = Ca, Mg):Mn^{4+} [357]. $BaTiO_3$:Ho^{3+}/Yb^{3+} [358], $BaTiO_3$:Er^{3+} [359], $BaTiO_3$:Eu [360], $BaTiO_3$:Er^{3+} [361], $BaTiO_3$:Nb^{3+}/SiO_2 [362], etc.

3.4 Sonochemical

Sonochemistry is considered to be a well-developed process for the synthesis of various nanomaterials since from early of 1980. A typical process involves ultrasound radiation (20 kHz–10 MHz) mediated via chemical reaction of the molecules. The acoustic cavitation is the main physical phenomenon liable for the sonochemical process. Acoustic cavitation including the formation, growth and implosive collapse of bubbles in a liquid provides exciting conditions (~5,000 K, pressures of ~1,000 bar, heating and cooling rates of >10^{10} K/s) inside the collapsing bubble which permits the origin of chemical reactions in liquids or liquid-solid slurries. This agrees for the preparation of a wide range of hierarchical nanostructured materials. In this technique, many external experimental parameters influence the chemical reactions, including temperature, intensity of ultrasound frequency and pressure. The temperature of the reaction precursors greatly affects the sonochemical rate which is mainly related to the composition of bubbles before collapse. Generally, greater the ambient temperature, lesser the sonochemical rate. However, static pressure influences few factors, namely nucleation of the cavity, solubility and collapse of the cavity. These factors act contrarily with increase of the static pressure. Successively, the intensity of the acoustic sound frequency mainly influences the rates of sonochemical reactions. Above the threshold value, the sonochemical rate will increase with better intensity of the sound field [363–365]. Undoped $BaTiO_3$ and $BaTiO_3$:RE (where RE=Sm^{3+}, Ce^{3+}, Dy^{3+}, Tb^{3+} and Eu^{3+}) nanophosphors were synthesized via sonochemical route [366] by using [$Ba(NO_3)_2$] (1.11 g, 4.25 mmol), [$Ti(OPr)_4$] (1.20 g, 4.25 mmol) and NH_4OH. The pH of the reaction solution was maintained at 9. The high intensity (100W/cm^2, 20kHz) ultrasound was irradiated to the reaction mixture under air for 1.5 h. At the end of the reaction, the precipitate was obtained and washed with water. Finally, the obtained white powder was calcined at 900°C for ~5 h. By addition of RE ions, similar experimental procedure was followed to prepare $BaTiO_3$:RE nanophosphors. They found that the prepared samples exhibit single-phased rod-like structures. Multi-color light emission under single excitation wavelength was clearly noticed. M. Dhanalakshmi et al. [367] reported that $BaTiO_3$:Dy^{3+} (1–5 mol%) nano-powder was synthesized via sonochemical route using barbituric acid. The body-centered cubic phase of the prepared samples was confirmed from PXRD results. The morphological changes with alteration of different experimental factors, namely sonication times and concentrations of the barbituric acid, was systematically studied. The PL emission spectra consisting of peaks at ~480, 575 and 637 nm was clearly noticed. The application of the prepared samples in effective detection of latent fingerprints and lip prints was systematically discussed. Dang et al. [368] synthesized $BaTiO_3$ nanoparticles in ethanol–water mixture solution by maintaining a low temperature. Upon ultrasound irradiation, various growth mechanisms of the $BaTiO_3$ particles in aqueous and ethanol–water solution were presented. They found that the ultrasound irradiation prevents the random aggregation of nanocrystals and oriented nanocrystals in specific crystal face by increasing the concentration of hydroxyl groups. The presented results clearly demonstrated that the ultrasonication plays a key role in oriented aggregation of nanocrystals. However, Yeshodamma et al. [369] studied structural and photoluminescence properties of $SrTiO_3$: Pr^{3+}: A^+ (A^+: Li, Na, K) nanophosphors synthesized via sonochemical route. The

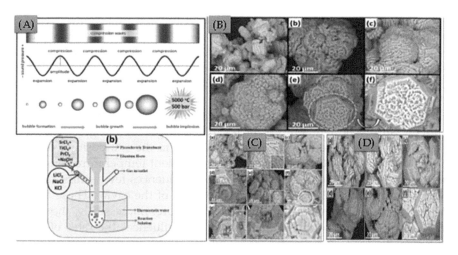

FIGURE 3.8
(a) Schematic diagram to show steps involved in the sonochemical synthesis of $SrTiO_3$: Pr^{3+} (1–11 mol%), SEM images of $SrTiO_3$: Pr^{3+} (5 mol%) NPs prepared by the (b) ultrasound irradiation time, (c) concentrations of biosurfactant EGCG, and (d) different pH levels. (Reprinted with the permission from Ref. [369] Copyright © Elsevier publications.)

growth mechanism of hierarchical structures was systematically studied (Figure 3.8a). Morphological changes corresponding to various experimental factors, such as surfactant EGCG concentration (5–35%W/V), sonication time (1–5 h), pH (1, 3, 5, 7 and 9) and sonication power (20–26 kHz), were systematically studied (Figure 3.8b–d). In addition, the role of EGCG as a face-inhibitor for the formation of lower-like hierarchical structures was well discussed. All superstructures were well organized with least utilization of energy as the nature prefers. The enhancement in luminescence intensity upon codoping of Li^+, Na^+ and K^+ ions in $SrTiO_3$:Pr^{3+} was clearly noticed. Prepared phosphors exhibit short lifetime, good quantum efficiency, and excellent color purity which is near to the NTSC standards.

Dhanalakshmi and coworkers [370] explored the superstructures of $BaTiO_3$:Eu^{3+} nanophosphors synthesized via sonochemical route using biotemplate *Aloe vera* (*A.V.*) gel as a surfactant (Figure 3.9a). The formation of biphasic (tetragonal and cubic) was confirmed by PXRD results. The role of concentration of *A.V.* gel and ultrasound irradiation time on superstructures formation was clearly discussed. The biotemplate *A.V.* gel plays a significant role for the formation of superstructures, which creates the system as anisotropic due to the presence of wrapped polymeric and protein chains in *A.V.* gel. Hence, smaller particles coalesce in specific oriented directions resulting in $BaTiO_3$:Eu^{3+}superstructures (Figure 3.9b–d). Some of the materials prepared via sonochemical route are $SrTiO_3$ [371], SiO_2@$SrTiO_3$:Dy^{3+} (5 mol%) [372], $SrTiO_3$@SiO_2 [373], $BaTiO_3$ [374], etc.

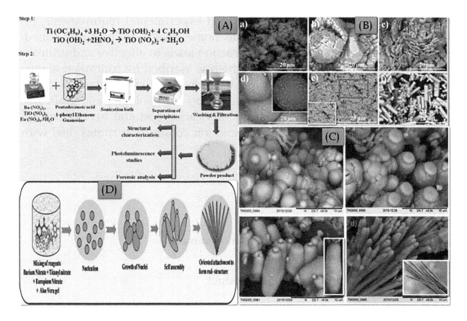

FIGURE 3.9
(a) Schematic for the sonochemical synthesis of $SrTiO_3$: Eu^{3+} (1–11 mol%), SEM images of $SrTiO_3$: Eu^{3+} (5 mol%) NPs prepared by the (b) concentrations of biosurfactant *A.V.* gel (5–35%W/V), (c) ultrasound irradiation time (1–4 H), and (d) Systematic illustration to show the formation mechanism based on self-assembly by oriented attachment of $BaTiO_3$: Eu^{3+}. (Reprinted with the permission from Ref. [370] Copyright © Elsevier publications.)

3.5 Solution Combustion

The Kingsley and his coworkers first developed the simple and rapid self-igniting solution combustion route to fabricate various nanomaterials [375]. The basic idea of the solution of combustion route was originated from the thermochemical reactions utilized in explosives and propellants. Accordingly, the stoichiometric calculations of the basic materials are estimated. During self-ignited combustion, the large amount of energy liberated from the exothermic reaction between precursor and fuels. This results in final fine powders by avoiding grain growth and agglomeration [376]. By selective basic materials with required stoichiometry, homogeneous and highly stable and pure materials can be easily fabricated in a simple, fast and without complicated experimental requirements of complicated experimental set-ups. The solution combustion classified it into four types (i) self-propagating high-temperature synthesis, (ii) flame synthesis, (iii) solution combustion and (iv) redox compounds combustion process [377,378]. The solution combustion route offers significant features, such as (i) simple, rapid process and comparatively no complicated experimental requirements, (ii) the stoichiometry, purity, structure and homogeneity can be easily tuned, (iii) highly exothermicity during the reaction leads to effective and uniform substitution of impurities into host, which is ascribed by controlled atomic mixing of reactants in the initial precursors, (iv) stabilized metastable states can be achieved by controlled reaction, and (v) morphologies and grain size of the particles can be tuned and low fabrication cost and hence may use large-scale productions. In addition, solution combustion routes deal with significant advantages, namely;

- It was rapid and single-step method.
- It was auto-igniting route.
- The reaction is exothermic.
- Comparatively less temperature requirement.
- It was low production cost.
- Homogeneity in the prepared product.

Although the solution combustion route has many advantages, it has some limitations, such as

- The fabricated samples have rough surface and more aggregation.
- More time consumption for fuel preparation and final product yield is less.
- The fuels used are hazardous and liberated gases are toxic.
- As reaction is exothermic, if the temperature is not controlled it may result explosion.

The solution combustion route influenced many experimental parameters, namely fuel, water content, flame temperature and gaseous products released during combustion.

3.5.1 Role of Fuels

The pivotal role of fuels in solution combustion route support burning the precursor solution by breaking C-H bonds. The ignition temperature of these fuels is less than that of actual phase transformation temperature and is combustible due to the presence of N-N bond that decomposes exothermically to N_2 (N≡N). The fuels also act as a facilitator to homogenous mixing of metal ions, eg, Carbohydrazide (CH), Oxalyl dihydrazide (ODH), Malonic dihydrazide (MDH), urea, glycine, etc.

3.5.2 Water Content

The water content during the combustion is an important aspect to achieve well crystalline nanomaterials. The excess water content may reduce the flame temperature which leads to incomplete burning of redox mixture and results in noncrystalline product. In addition, the surface area also decreases with increase in water content.

3.5.3 Importance of Gaseous Products

The gases liberated during combustion (H_2O, CO_2 and N_2, etc.) may diminish the heat generation. These gaseous products released during combustion help in dissipating the heat released during combustion reaction. The increased gas liberation results in increasing the surface area with micro- and nanoporous structures.

3.5.4 Flame Temperature

The flame temperature plays a major role in solution combustion to desired product. The flame temperature mainly depends on decomposition of fuels and liberated gaseous products during exothermic reaction. The increased liberation of gases results in decreased flame temperature.

3.5.5 Fuel-to-Oxidizer Ratio

The fuel-to-oxidizer ratio is a key parameter in solution combustion route. The uniform combustion takes place only when the fuel-to-oxidizer ratio is maintained unity or greater than unity. Combustion may not take place at all if the stoichiometry is not maintained. High-quality crystalline nanomaterials can be achieved by increasing the fuel ratio, hence no further calcination of product is required.

In 2019, Yin and coworkers [379] demonstrated the synthesis of $Ca(Ti_{1-x}Zr_x)O_3:Dy^{3+}$ phosphors via combustion route. The crystal distortion of $CaTiO_3$ after incorporation of Zr^{4+} was confirmed via Rietveld refinement of the XRD profiles. The change in orthorhombic perovskite phase of $CaTiO_3$ to $CaZrO_3$ phase while increasing the concentration of Zr^{4+} was investigated. The substantial PL emission intensity enhancement upon codoping of Zr^{4+} was noticed. The quantum efficiency of the prepared phosphor was found to be 9.9%, and lifetime diminished with the increasing codoping of Zr^{4+} concentration. The nanometric grains and elemental composition of the products were confirmed through SEM and EDS, respectively. The photometric coordinates of the prepared products were closest to the standard white-light color coordinate and hence present phosphor was considered to be a potential material for application in white light-emitting diodes. This preparation route was further used for the synthesis of europium-doped (1–9 mol%) $SrTiO_3$ nanophosphor using urea as a fuel by Bhagya et al. [380]. Strontium nitrate [$(Sr(NO_3)_2$, 99%], tetra-n-butyl titanate [$Ti(OC_2H_9)_4$, 99%], urea [NH_2CONH_2, 99%], 1:1 nitric acid [HNO_3, 99%] and europium oxide [(Eu_2O_3), 99%] were used as initial materials and without any further purification. The synthesis procedure involves preparation of titanyl nitrate by using tetra-n-butyl titanate with addition of distilled water and 1:1 HNO_3. The following reactions take place during the formation:

$$Ti(OC_4H_9)+3H_2O \rightarrow \circledR TiO(OH)_2 +4C_4H_9OH$$

$$TiO(OH)_2 +2HNO_3 \rightarrow TiO(NO_3)_2 +2H_2O$$

The stoichiometric quantities of obtained titanyl nitrate solution, strontium nitrate, europium nitrate [dissolved europium oxide in 1:1 HNO_3 in sand bath] and urea were well mixed in distilled water using a magnetic stirrer. The obtained precursor solution is placed in preheated muffle furnace maintained at $450°C \pm 10°C$. Dehydration of the mixture was noticed with the liberation of a large amount of gases and the solution boiled. A flame appeared on the surface of the froth and continued quickly throughout the whole volume, leaving a white powder. The as- formed powder was calcined at 600°C for 2h. The PXRD profiles of the samples confirm cubic phase and were matched with JCPDS card no: 35–0734 with space group Pm–3m (No-221). The porous and agglomeration nature of the samples was confirmed from FE-SEM images. The PL emission spectra shows characteristic peaks arise from $^5D_0 \rightarrow {}^7F_{1,2,3}$ intra-4f shell Eu^{3+} ion transitions. An intense orange red emission with high color purity was noticed from photometric studies. Sandhyarani et al. [381] synthesized $SiO_2@SrTiO_3:Eu^{3+}$ (1 mol%), Li^+ (1 wt%) core–shell nanopowders via solution combustion route to encounter visualization quality enhancement of latent finger prints by traditional powder dusting process. Firstly, typical synthesis procedure involves preparation of micrometer-sized uniform SiO_2 spheres followed by well-known Stöber method. Later, core–shell-structured $SiO_2@SrTiO_3:Eu^{3+}$ (1 mol%), Li^+ (1 wt%) nanopowders were synthesized via solution combustion method. The systematic formation mechanism of core–shell nanopowders was illustrated (Figure 3.10a). The PXRD profiles exhibiting a broad peak at $2\theta=22.8°$ belong to amorphous SiO_2 (JCPDS card No.29–0085) and other sharp peaks reveal the cubic phase of $SrTiO_3$ (JCPDS card No.35-0734). The uniform size with spherical-shaped core–shell particles was revealed from SEM as well as PXRD studies. The CIE chromaticity coordinates of prepared nanopowders tuned from orange red to the deep red with increase of coating up to four cycles. The two-fold PL intensity enhancement was noticed upon four cycles

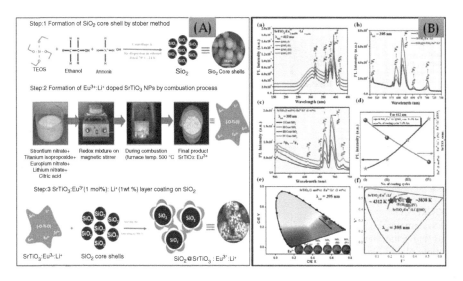

FIGURE 3.10

(a) Step-by-step formation mechanism of prepared core–shell nanopowders and (b). PL excitation spectrum, emission spectra under 395 nm excitation, emission spectra of SiO$_2$@SrTiO$_3$:
Eu^{3+}(1 mol%), Li$^+$ (1 wt%) nanopowders with different coat (1–6), variation of PL intensity
with respective to SrTiO$_3$ concentration & number of coating cycles, CIE and CCT diagram.
(Reprinted with the permission from Ref. [381] Copyright © Elsevier publications.)

of coated samples (Figure 3.10b). In 2014, Shivram and group [382] demonstrated the synthesis of orthorhombic-structured CaTiO$_3$:Eu^{3+} (1–9 mol%)
phosphors via low temperature (500°C) solution combustion route using
urea as a fuel. The crystallite size of the prepared samples was found to be
in the range of 40–45 nm. From SEM images, foamy and porous with large
agglomerates was noticed. The characteristic PL emission peaks noticed at
~540, 593, 615, 653, 696 and 706 nm ($^5D_0 \rightarrow {}^7F_{j=0,1,2,3,4,5}$) were attributed to Eu^{3+}
ions. In addition, the prepared phosphor showed CIE chromaticity coordi-
nates (x, y) very close to NTSC standard value of red emission, as a result it
was quite useful for display applications.

Recently, Inamdar and coworkers [383] successfully fabricated CaTiO$_3$:Eu^{3+}
luminescent nanoparticles and polypyrrole/CaTiO$_3$:Eu^{3+} nanocomposites
via solution combustion and solution casting techniques, respectively. The
systematic illustration for the fabrication of the polypyrrole/CaTiO$_3$:Eu^{3+}
nanocomposites was demonstrated (Figure 3.11a). The single orthorhombic
phase and crystallite size of ~40–45 nm was estimated from PXRD results.
The CaTiO$_3$:Eu^{3+} deposition on the polypyrrole was clearly noticed in SEM
images. The prepared phosphors displayed splendid red emission at 398 nm
excitation. The optimized concentration of the Eu^{3+} ions has been found to
be 7 mol% in polypyrrole polymer matrix. The average CCT value was found
to be ~2,234 K (Figure 3.11b), indicating that the product is highly useful

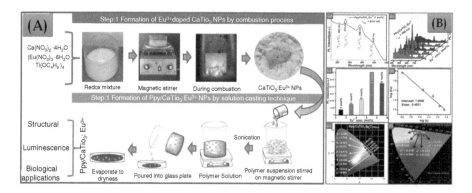

FIGURE 3.11

(a) Systematic illustration for the fabrication of the polypyrrole/ $CaTiO_3:Eu^{3+}$ nanocomposites and (b) Excitation spectrum, emission spectra, variation graph of PL intensity vs Eu^{3+} concentration, plot of log (x) vs log (I/x) and CIE, CCT diagrams of polypyrrole/$CaTiO_3$: Eu^{3+} nanocomposites. (Reprinted with the permission from Ref. [383] Copyright © Elsevier publications.)

in warm LEDs. Some of the other titanium-centered perovskites prepared via solution combustion route are $MTiO_3$ (M=Ca or Sr): Eu^{3+} or Pr^{3+} [384], $BaTiO_3:Eu^{3+}$ [385], $SrTiO_3$ [386], $CaTiO_3$ [387], $CaTiO_3:Sm^{3+}$ [388], etc.

3.6 Solid-State Reaction

The solid-state reaction route is commonly known as conventional mixed oxide method, which is extensively used for the preparation of wide range of oxides, (oxy) fluorides, (oxy) nitrides as well as (oxy) chlorides. The typical synthesis route, involving stoichiometric amount of individual solid precursors, is thoroughly mixed by grinding to obtain homogeneous mixture [389]. The grinding process convinces that all precursors have been well mixed which results in increase of surface area. Normally, various methods, such as mortar and pestle, ball milling, etc. were employed for grinding process. The grounded powder is usually treated with heat at high temperatures for extended time to enable diffusion of atoms or ions in the chemical reaction. The high-temperature annealing of the powder leads to improved crystallinity with extensive particle size distribution [390,391]. In addition, thermal reactivity as well as compaction of powder considerably enhances by reducing crystallite size and forming lattice defects in the powder. Generally, wet grinding was more preferable than dry grinding which adversely impact on crystallite size. Even though conventional solid-state reaction process has

TABLE 3.2

List of Advantages and Limitations Offered by Solid-State Reaction Route as Compared with Other Routes

Synthesis Routes	Advantages	Limitations
Solid-state reaction	Simple and rapid, crystalline nature of the product, high yields, cheap raw materials, process control	High energy consumption, high temperatures needed
Hydrothermal	Simple, good control over different morphologies, small particle size, uniform particle distribution	High pressure needed, special autoclaves required
Sonochemical	Homogeneous product, process control, low cost; easy operation	Long duration, heat treatment needed

some limitations, but still in practice to synthesis various materials as compared with other synthesis routes. The advantages and limitations offered by solid-state reaction route as compared with hydrothermal, sol–gel and sonochemical methods are tabulated in Table 3.2.

Li et al. [392] prepared $CaTiO_3:Dy^{3+}/Eu^{3+}$ phosphors by a solid-state reaction. The $CaCO_3$ (99.9%), TiO_2 (99.9%), Eu_2O_3 (99.99%) and Dy_2O_3 (99.99%) were used as starting materials. The effective substitution of Dy^{3+} and Eu^{3+} in a $CaTiO_3$ host was confirmed through XRD results. The spectral overlap of emission and excitation spectrum of Dy^{3+} and Eu^{3+}, respectively, confirmed the possible energy transfer phenomena between Dy^{3+} and Eu^{3+} ions. This energy transfer results in luminescence tuning, which offers designing of the WLEDs. Luminescent $CaTiO_3:Eu^{3+}$ red phosphors were synthesized using H_3BO_3-assisted solid-state method by Panpan and coworkers [393]. Analytical reagent (AR) $CaCO_3$, TiO_2, H_3BO_3 and Eu_2O_3 (99.99%) were used as initial materials without further purification. Sharp PXRD peaks confirm the orthorhombic phase of $CaTiO_3$ and are well agreement with JCPDS card No. 76-2400. Similarly, Tian et al. [394] synthesized H_3BO_3 flux-assisted $CaTiO_3$: Pr^{3+} red phosphor. The PXRD results of sample reveals the orthorhombic perovskite phase without any impurity. The CIE chromaticity coordinates were very close to that of the ideal red light. Following this work, the synthesis of red-emitting $CaTiO_3:Pr^{3+}$ phosphor with enhanced PL and ultrahigh temperature-sensing properties via NaF flux-assisted solid-state reaction was demonstrated by the same research group [395]. This method was further employed for the preparation of Er^{3+}/Yb^{3+} doped $SrTiO_3$ ceramic phosphor using $SrCO_3$, TiO_2, Yb_2O_3 and Er_2O_3 as raw materials by Sukul et al. [396]. The 20-fold enhancement of up-conversion emission intensity was noticed owing to the alteration of cubic-to-tetragonal phase. In 2015,

TABLE 3.3

Information on Selected Titanium-Centered ABO_3 Perovskites Synthesized via Solid-State Reaction Method

Product	Temperature/ Time (°C/h)	Comments	References
$CaTiO_3$:Pr	900/2	• Improved luminescent properties of the phosphor, boric acid as flux regent and aluminum ion as charge compensator were added. • The effect of H_3BO_3 does not only control the mean size, size distribution and shape of the phosphor particles, but also participates in the formation of suitable lattice defects of matrix.	Chen et al. [400]
$CaTiO_3$:Pr^{3+}	1200/3	• The presence of rare earths reduced the intensity of red emission at 612 nm except in case of Gd^{3+} which enhanced the PL emission by 25% and afterglow emission intensity by 10% more. • Codoping of Eu^{3+} produced an additional absorption peak (466 nm) in visible region without suppressing the UV bands.	Panigrahi et al. [401]
$CaTiO_3$:Cr^{3+}	1350	• Exhibit near-infrared afterglow, due to the creation of new defects. • Successfully break the monopoly of the ordinary Cr^{3+}-activated gallate phosphors with the spinel structure.	Qin et al. [402]
$CaTiO_3$:Eu^{3+}	1200	• The effects of calcination condition (temperature, duration, and Eu dopant concentration) on the structural and Eu^{3+} emission properties are studied in detail. • Temperature dependences of the PL intensity and decay time at T=20–300 K are measured and analyzed on the basis of the thermal quenching model.	Orihashi et al. [403]
$BaTiO_3$	1200/2	• The changes in surface charge properties of the film with respect to cysteine concentrations were determined using a current–voltage (I-V) technique. • The current response increased with cysteine concentration (linear concentration range=10 μM–1 mM).	Selvarajan et al. [404]

(Continued)

TABLE 3.3 (*Continued*)

Information on Selected Titanium-Centered ABO_3 Perovskites Synthesized via Solid-State Reaction Method

Product	Temperature/ Time (°C/h)	Comments	References
BaTiO$_3$:Sm^{3+}	1100/5	• The phosphors can be effectively excited by the near-ultraviolet light 409 nm produces the emitted reddish orange light peaks locate at about 561, 595 and 643 nm.	Zhang et al. [405]
BaTiO$_3$	1200/2	• The simple, novel device structure leads to new dimensions in diagnosis, namely piezoelectric-based biosensing. • The piezoelectric output signal from the device contains the biosensing signal, thus having a dual purpose as a generator and biosensor.	Selvarajan et al. [406]
BaTiO$_3$:Eu^{3+}	1200/3	• An intense and sharp emission peak at 615 nm was exhibited by the phosphors upon excitation at 397 nm. • The incorporation of K$^+$ ions in optimized Ba$_{0.95}$TiO$_3$:0.05 Eu^{3+} phosphor resulted in a remarkable enhancement of photoluminescence intensity by 2.33 times as compared to bare one.	Dhananjay et al. [407]
BaTiO$_3$: Er^{3+}	1200/3	• Phosphors exhibited efficient dual emission mode with a sharp and intense visible green emission peaks centered at 525 and 550 nm via the up-conversion process by using 980 nm Laser source excitation and the down-conversion process upon the 380 nm UV light excitation.	Dhananjay et al. [408]
CaTiO$_3$:Eu^{3+}		• PL data showed that this phosphor was efficiently excited by near-ultraviolet (NUV) light at wavelength around 400 nm and emitted intense red light with a broad peak around 618 nm. • The concentration quenching and thermal quenching of the samples are discussed.	Fu et al. [409]

(*Continued*)

TABLE 3.3 (*Continued*)

Information on Selected Titanium-Centered ABO_3 Perovskites Synthesized via Solid-State Reaction Method

Product	Temperature/ Time (°C/h)	Comments	References
$CaTiO_3$:Pr	600–1000/2	• PL emission spectrum consists of a very intense single red band, with a maximum at about 614 nm, corresponding to the characteristic $^1D_2 \rightarrow {}^3H_4$ transition of Pr^{3+}. • The afterglow in $CaTiO_3$:Pr contains a bright but short (less than a few minutes) component followed by a weak but long persistent tail (around 20 min).	Yin et al. [410]
$CaTiO_3$:Pr^{3+}	1200/4	• The concentration of In^{3+} was varied from 0.05 to 0.3 mol% and it was found that the In^{3+} incorporation enhanced the intensity of the red emission and the afterglow decay time of the $CaTiO_3$:Pr^{3+} phosphor considerably. • Incorporated In^{3+} charge compensators enhanced the luminescence intensity of the phosphor by neutralizing the additional positive charge generated by the Pr^{3+} ions when substituting in the Ca^{2+} ions site.	Noto et al. [411]
$SrTiO_3$:Pr^{3+}	1300/4	• The enhanced red luminescence was observed for the doubly doped compositions compared to the host matrix with only Pr^{3+} as an activator. • Host to activator energy transfer seems to be responsible for this efficiency enhancement.	Ryu et al. [412]

Zhang et al. [397] demonstrated an ecofriendly $NaCl$–H_2O-assisted synthesis of $SrTiO_3$ nanoparticles by solid-state reaction at low temperature from commercially available raw materials ($SrCO_3$ and rutile). This work clearly demonstrated the role of both $NaCl$ and H_2O in order to accelerate the formation of $SrTiO_3$ nanoparticles at relatively low temperature. Based on obtained experimental results, a rational growth mechanism was recommended as well as discussed in detail. A series of red-emitting $CaTiO_3$:Sm^{3+} phosphors prepared via solid-state reaction method was by Ha et al. [398]. In that work, the monoclinic $CaTiO_3$ phase was clearly noticed. SEM results reveal uniform structure with micrometer grain size. An excellent red–orange light emission owing to $^4G_{5/2} \rightarrow {}^6H_{7/2}$ transition of Sm^{3+} ions upon near-ultraviolet excitation was clearly noticed. Jyothi et al. [399] demonstrated the regulation of charge transfer transitions in $SrTiO_3$:Y^{3+}, Eu^{3+} red phosphors via site selective replacement and leads to improved emission intensity with high color purity. A 14-times enhancement in PL emission intensity was monitored by selective substitution of Y^{3+} ions at Sr^{2+} and Ti^{4+} sites in $Sr_{0.95}Eu_{0.05}TiO_3$, which due to increased asymmetric ratio between transitions of Eu^{3+} ions. In addition, dual Eu^{3+} and Y^{3+} ions substitution in the host site yields PL emission, which is close similarity with that of the commercial red phosphor. Further, the list of titanium-based perovskites synthesized by solid-state reaction route were tabulated in Table 3.3.

3.7 Pulsed Laser Deposition

From past decades, the lasers play a significant role in various areas, namely medical, metallurgy, electronics, materials research, etc. due to its excellent coherence and extraordinary monochromatic nature. The lasers act as both passive and active tool in process monitoring system as well as material process coupled with its energy [413]. For example, laser melting, etching and ablation, laser-induced hardening, novel phase formations and laser deposition to achieve thin films. The pulsed laser deposition technique is a simplistic and adaptable fabrication route among all other thin film growth techniques [414]. Generally, it consists of a target and substrate holders evacuated in the vacuum chamber. The typical fabrication process involves the interaction of high power laser light with target surface through focusing optics [415]. Due to high energy interaction, the surface of the target material melts and vaporizes. This vapor deposited on the suitable material placed in front of the target results in excellent quality of thin film with same parent stoichiometry. Normally, pulsed laser deposition can be classified into three types based on the type of interaction between laser beam and target. They are (i) laser beam and target interaction results to vaporization, (ii) interaction between laser beam and target material leads to transport of the vapor plume, and (iii) anisotropic adiabatic expansion of the plasma-induced film formation. The major advantages of pulsed laser deposition are;

- Versatile technique: Materials can be deposited on the target surfaces without altering the stoichiometry of the parent, multilayer deposition.
- Economic: Single laser can use for many vacuum systems.
- Rapid: High deposition rates, typically ~100s Å/min.
- Scalable: as complex oxides move toward volume production.

In spite of these significant advantages, several reasons confined the use of pulsed laser deposition techniques, namely;

- The plasma plume generated at the time of the laser ablation process is highly forward directed. Hence, thickness of the deposited film is nonuniform.
- The deposited material generally comprises macroscopic bubbles of molten material. These bubbles unfavorable to the properties of the ablated thin film.
- The fundamental processes involved in the laser-generated plasmas are not fully understood.

In 2010, Yang et al. [416] obtained Li-doped $CaTiO_3:Pr^{3+}$ thin films ablated on Si (100) substrate via pulsed laser deposition technique. The stoichiometric quantities of $CaCO_3$ (99.9%), Pr_2O_3 (99.99%), TiO_2 (99.9%) and LiF (99. 9%) were used as starting materials. The thin film was grown by using a pulsed laser (KrF excimer, ~248 nm) with flux of ~ 3.5 J/cm^2 as well as maintained repetition rate was 5 Hz. The target and Si (100) substrate was separated at a distance of 35 mm. Xiong et al. [417] demonstrated the self-organization of Ni nanocrystals ablated on the epitaxial $SrTiO_3$ matrix via pulsed laser deposition route. The matrix and nanocrystals were specifically altered by monitoring the growth conditions. The possible growth transformation mechanism from 2D layer-by-layer $SrTiO_3$ to 3D islands Ni nanocrystals was discussed in detail. The orientations of $SrTiO_3$ plane (0 0 1) with Ni (1 1 1) were clearly revealed. The excimer laser-assisted $CaTiO_3:Pr$ phosphor film at room temperature was demonstrated by Tsuchiya and coworkers [418]. The effects of the substrate materials and laser fluence on photoluminescence were discussed in detail. The formation mechanism of KrF laser-assisted ablation of $CaTiO_3:Pr$ film was established based on thermal simulations as well as atomic force microscopy results. Fasasi et al. [419] reported $BaTiO_3:Gd$ thin films fabricated by laser deposition method on glass and silicon substrates using Gd_2O_3 (99.9%), TiO_2 (99.9%) and BaO (97%) as starting materials. The effective substitution of Gd in Ba lattice leading to oxygen vacancy compensation with structural modification and an average grain size of 30 nm was noticed.

3.8 Sputtering Method

Sputtering is considered to be most significant technique in physical vapor deposition, which was first developed in 1970. Till date, many sputter techniques are available for the fabrication of nanomaterials, such as radio frequency, reactive, ion beam, magnetron sputtering and diode [420]. This method comprises of atoms sputter from a target material due to interaction of the target material and energetic ions. These sputtered atoms have specific energy and orientation were accelerated and deposited on the surface of the substrate in high pressure results formation of thin films [421]. Normally, the rate of atoms sputter from the target mainly depends on the number of bombarding ions hitting the target, composition of target material, properties of the incident ions, experimental geometry as well as binding energy. This method offers several advantages, including greater uniformity, density and interface roughness of the deposited film, deposition can be possible over large surface area and high melting point materials can also be deposited [422,423]. Hence, sputtering processes have become one of the versatile methods for preparing high-quality thin solid films of any material. Sarakha et al. [424] reported the fabrication of $CaTiO_3:Pr^{3+}$ thin films via radio frequency sputtering technique in the presence of pure argon gas. The variation of electrical and optical properties of the thin films with deposition pressure (0.125–4.5 Pa) as well as annealing parameters were discussed in detail. In 2005, $CaTiO_3:Pr$ red phosphor thin films were deposited on ITO/glass substrates, ZnO/ITO/glass as well as glass via RF magnetron sputtering by Chung and coworkers [425]. The influence of various substrates and heat treatment on the structural and luminous properties was discussed. They obtained improved luminescent properties of the prepared films by post heat treatment, which owing to phase transformation from amorphous to poly crystalline as well as abolition of microdefects. Recently, Shihab and coworkers [426] reported the fabrication of 1-D photonic $BaTiO_3$ crystal via RF sputtering technique for optical field-induced nonlinear absorption and PL properties. From FESEM images, uniformity in deposited layer thickness was clearly noticed. This method was further employed for the fabrication of Er-doped $BaTiO_3$ thin films via RF magnetron sputtering method [427]. A detailed study on effect of dopant Er concentration on the structural, morphological and optical properties of $BaTiO_3$ films was performed. A larger spherical grains with neck formation was confirmed from SEM images. The substitution of Er^{3+} ions in Ba (A-site) and Ti (B-site) of $BaTiO_3$ was clearly revealed.

4

Titanate Materials

Titanium-centered ABO_3 structured perovskites offers several exciting properties, namely dielectric [302,314,428–431], thermal [430,432–434], optical [435–437], ferroelectric [438,439], piezo-electric [440,441], electrical [442–444], superconducting [445,446], etc. due to their distorted cationic configuration, non-stoichiometric cations or anions, mixed valence electronic structures, stability of metallic-based ABO_3 structures, and chances of fabricating multi-component structures via effective replacement of cations in A and B sites [447]. These outstanding properties account for many significant applications, namely sensors and actuators, memory devices, capacitors, higher temperature heating devices, frequency filters, etc. [448–455]. Few perovskites materials offering the aforementioned properties are discussed in detail below.

4.1 BaTiO$_3$ Multipods

Dielectric nature is an inherent property of a material that can inhibit the movement of electron and therefore induce polarization within the material upon applied external electric field. Dielectric materials possess higher resistance to electric current upon applied dc voltage and hence small deviations in the electrical properties as compared to conducting materials [456]. Dielectric properties such as dissipation factors, dielectric constant (ε), dielectric resistivity, and dielectric strength are utilized to discuss the dielectric nature of a material. The dielectric constant is the ratio of permittivity of a substance to the permittivity of a free space. Usually, dielectric constant mainly depends on several factors, such as (i) uniformity and thickness of a material, (ii) contact lacking between electrodes and samples, (iii) adsorption of water, and (iv) pinholes, contact resistance, and edge effects [457]. However, dielectric strength of the materials highly depends on homogeneity of a material, shape of electrodes, geometry of a specimen, stress mode (either AC, DC, or pulsed), and ambient environment. Therefore, continuous efforts are made to enhance the dielectric properties of materials. Among various materials, titanium-based perovskites are considered to be most promising candidates due to least dielectric loss and higher dielectric constant [458]. A generic example for excellent dielectric material is $BaTiO_3$, which is the

DOI: 10.1201/9781003381907-4

first fabricated piezoelectric material. The crystallographic lattice size of BaTiO$_3$ varies with temperature owing to distorted octahedral TiO$_6$ as the temperature decreases from higher to lower temperature [459,460]. In addition, large spontaneous polarization induced by distorted octahedral results in increased dielectric constant. At temperatures greater than Curie temperature (120°C), BaTiO$_3$ organizes all the Ba^{2+} ions and occupies eight corners of cubic cell; however single Ti^{4+} and O^{2-} ions occupy the center of each surface of the cube. Below the Curie temperature BaTiO$_3$ exists in the distorted tetragonal crystal structure with a mutual dislocation of the centers of both positive and negative charges within the sublattice [461,462]. As a result, a dipole moment is generated parallel to one area of the cubic phase. This spontaneous polarization created in the tetragonal structure is the cause of the piezoelectric and ferroelectric behavior of the sample [463,464]. In 2014, Nayak et al. obtained perovskite BaTiO$_3$ multipods by solid-state reaction route [465]. The powder X-ray diffraction (PXRD) results show traces of diverse impurity peaks in ethanol- (mentioned as 24 E) and HCl (24 H)-media-washed BaTiO$_3$ compared with purified BaTiO$_3$ samples. The sharp diffraction peaks of purified BaTiO$_3$ powder revealed tetragonal crystal structure and were well matched with standard JCPDS card no. 79–2264 [466]. In addition, PXRD profile splitting at $2\theta=45°$ was clearly noticed and it further confirms the tetragonal structure of the prepared sample [467]. From field emission scanning elctron microscope (FE-SEM) images, spherical particles of pure titania was clearly noticed. However, Ba(OH)$_2$.8H$_2$O particles exhibit rice grain-shaped crystals. Various phases of sample which depend on calcination temperature was discussed in detail. At 700°C calcination, clustered mass with few two rod-like structured particles which owing to inter-diffusion of counter ions was clearly noticed. At higher calcination temperature, the degree of inter-diffusion between Ti^{4+} and Ba^{2+} ions upsurges and results in increased concentration of BaTiO$_3$ nanoparticles by suppressing unreacted masses. The growth of BaTiO$_3$ particles in allied orientation at 1,000°C calcination results in the formation of grain-shaped particles. When the sample was calcined at 1,200°C, self-assembly and orientation process continues to form star-shaped BaTiO$_3$ multipods with average particle length and diameter of ~3 mm and 300 nm, respectively. The obtained results clearly reveal that the anisotropic growth of BaTiO$_3$ multipods can be engineered by controlling the reaction time and temperature [468]. The schematic diagram to demonstrate the formation process of BaTiO$_3$ multipods by inter-diffusion of Ti^{4+} and Ba^{2+} ions was clearly discussed (Figure 4.5c). As calcination temperature increases, rate and degree of inter-diffusion process also increases by releasing water vapors to form BaTiO$_3$ shell. This process continuous till the formation of thick BaTiO$_3$ shell by exhausts unreacted masses. At the highest calcination temperature, clear triangular blade-shaped (multipods) BaTiO$_3$ particles are formed which are clustered in the form of star.

The variation of dielectric constant (ε') and dielectric loss (ε'') versus frequency was systematically discussed. It was found that the small incremental changes of both ε' and ε'' with the fall in frequency in the range 10^6 to 10^2 Hz. When there is additional diminution in frequency, rapid increases of both ε' and ε'' were clearly noticed. The obtained result clearly indicates that the dielectric constant (ε') and dielectric loss (ε'') are independent of frequency over 10^3 to 10^6 Hz. Fewer dielectric loss (ε'') than dielectric constant (ε') over complete frequency range was observed. These results clearly reveal that the synthesized $BaTiO_3$ is considered as an ideal material for electronic device fabrication. Bode plot (AC impedance ($|Z|$) and phase angle versus frequency) of $BaTiO_3$ multipods were systematically studied. The increment in the impedance ($|Z|$) with the decrease in frequency was clearly noticed. The obtained outcomes clearly indicate the occurrence of space charge polarization in the sample. In addition, Nyquist plot (Z' versus Z'') shows semicircular arc for $BaTiO_3$ multipods, which is due to the existence of dielectric relaxation spectra with a shortrange relaxation time [469]. The Nyquist plot is characterized in the form of semi-circular arcs, which reveal the existence of narrow dielectric relaxation time in the spectra. The zero phase angle is both very low and high frequency range. Conversely, phase angle reaches $-90°$ at intermediary frequencies, which is the ideal angle for capacitor [470]. Generally, the dielectric loss cannot be accounted only by means of loss of capacitor, but it is a blend of capacitor and resistor. Various combinations of resistors and capacitors are conceivable to effectively characterize dielectric materials. Therefore, an equivalent circuit constituent of combination of capacitor and resistor was proposed in detail using electrochemical impedance spectroscopy data analysis software. The impedance of the parallel combination of RC circuit was given as follows [465]:

$$Z = R_g + \left[\frac{1}{1/R_{gb} + jwC_{gb}} \right]$$

From the obtained results, it was clearly evinced that the prepared $BaTiO_3$ multipods were considered as two-phased electrical systems i.e., one intergrain and another inside bulk. In addition, comparison was made on the dielectric property of $BaTiO_3$ multipods with TiO_2 and $Ba(OH)_2.8H_2O$ samples. The experimental results clearly demonstrated that the $BaTiO_3$ multipods show significantly greater dielectric constant as compared with commercial $BaTiO_3$ particles over frequency range from 10 Hz to 1 MHz. The slope of the Bode plot of $BaTiO_3$ multipods and commercial $BaTiO_3$ powder was found to be -0.844, which is very close to -1. Hence, it is clearly demonstrated that the prepared $BaTiO_3$ multipods can act as an ideal capacitor. However, plots of temperature-dependent dielectric constant of $BaTiO_3$ multipods at three frequencies (10, 10^3 and 10^6 Hz) reveals maximum and minimum dielectric constant at lowest (10 Hz) and highest frequency (1 MHz) measurements. The obtained greater dielectric constant is mainly owing to enhanced involvement of space charge polarization along with dipolar polarization.

The variation plot of impedance ($|Z|$) versus frequency (log f) at different temperatures was also presented. The increment in impedance ($|Z|$) with the decrease in frequency was clearly noticed, whereas $|Z|$ decreases with increase in temperature at any constant frequency.

4.2 PVDF-HFP-GMA/BaTiO₃ Nanocomposites

Energy storage materials are creating much interest for research community due to its high demand in modern electric power supply and renewable energy system [471–473]. The high dielectric materials, such as lead zirconium titanate and barium titanate, suffer from limitations, namely low breakdown strength and poor processing capacity [474]. Conversely, polymers possess high breakdown strength and excellent process ability, but show low dielectric constant. Hence, the introduction of inorganic ceramic fillers into insulating polymer matrix for fabricating high dielectric constant nanocomposites creates new avenue in energy allied applications [475–478]. Xie and group developed a PVDF-HFP-GMA/BaTiO₃ nanocomposite by utilizing covalent bonding between the functionalized nanoparticle and the modified polymer matrix to improve particle dispersion [479]. The poly (vinylidene fluoride-cohexafluoropropylene) [PVDF-HFP] was functionalized with glycidyl methacrylate (GMA) via atom transfer radical polymerization (ATRP). The amino-terminated silane molecules were used for surface engineering of BaTiO₃ nanoparticles. Further, the PVDF-HFP-GMA/BaTiO₃ nanocomposites were fabricated by grafting functionalized PVDF-HFP to the surface-modified BaTiO₃ nanoparticles via "grafting to" technique. By varying the BaTiO₃ nanoparticle volume fractions (10%, 20%, 30%, 40%, and 50%) in polymer matrix, many nanocomposites were fabricated and were indicated by GMA-BT-10, GMA-BT-20, GMA-BT-30, GMA-BT-40, and GMA-BT-50. The without-modified PVDF-HFP/BT nanocomposites were also prepared via the same blending method and were represented as BT-10, BT-20, BT-30, BT-40, and BT-50 for comparative study. The schematic illustration to demonstrate the fabrication process of the prepared nanocomposites was proposed. The typical fabrication process involves introduction of significant several active epoxy groups to PVDF-HFP by fluorine atom, which initiated the ATRP of the GMA monomers [480]. The dielectric properties of the PVDF-HFP-GMA/BT nanocomposites as a function of frequency at room temperature was systematically studied. The results clearly reveal the increment in the dielectric constant of the PVDF-HFP-GMA as compared with pure PVDF-HFP, which may be because of introduction of polar functional groups [481]. The remarkable increases of dielectric constant of the prepared nanocomposites with filling of BaTiO₃ nanoparticles were clearly noted. Converse results of the decreased dielectric loss with the increase of BaTiO₃ loading were obtained.

This decrement was mainly due the highly restricted polymer chain motion in the composites, which exhibit the show the advantage of the "grafting to" method. In addition, variation in electrical conductivity of the PVDF-HFP-GMA/BT nanocomposites with frequency was systematically studied. It was observed that the loading of BaTiO$_3$ nanoparticles slightly increases the electrical conductivity of the nanocomposites.

In addition, temperature-dependent dielectric properties of the PVDF-HFP-GMA/BT nanocomposites, the PVDF-HFP-GMA, and the PVDF-HFP are studied in detail (Figure 4.8). The obtained results reveal that the PVDF-HFP-GMA exhibit shows a strong dielectric response, especially at lower frequency ranges as compared to PVDF-HFP. Surprisingly, weaker temperature-dependent dielectric response was noticed when BT-APS nanoparticles were loaded in the polymer matrix. The obtained result clearly demonstrated the presence of covalent bonding between the nanoparticles and polymer matrix. This strong temperature-dependent dielectric behavior attributed from the electrical conduction was confirmed through temperature-dependent electrical conductivity of the samples. The PVDF-HFP-GMA exhibit frequency-independent greater electrical conductivity at high temperatures and low frequencies as compared with PVDF-HFP and the PVDF-HFP-GMA/BT nanocomposites was noticed. At higher temperatures, there is enhancement in the rotation of dipoles and mobility of charge carriers due to the presence of epoxy groups in the poly(GMA) brushes, which in turn leads to significant increment in the electrical conductivity as well as interfacial polarization [482]. As a result, the PVDF-HFP-GMA exhibit shows greater dielectric constants at lower frequencies and high temperatures. Conversely, the plot of dielectric loss factor versus frequency for the PVDF-HFP-GMA exhibit at high temperatures and low frequencies was found to be linear. When BaTiO$_3$ nanoparticles were loaded, the amino groups reacted with epoxy groups present in the GMA, which reduced the epoxy groups, resulting in covalent bonding between the BaTiO$_3$ particles and the polymer matrix. This suppressed the charge transport in the polymer matrix which led to weak frequency-dependent dielectric response, less dielectric loss, as well as lower conductivity at elevated temperatures and low frequencies, when compared with PVDF-HFP-GMA exhibit. These obtained features are favorable to potential energy storage and thermal management applications.

4.3 BaTiO$_3$/Polyvinylidene Fluoride (BT/PVDF) Nanocomposites

The surface engineering of BaTiO$_3$ particles and size reliant polarization are of technological significance in fabricating improved BT/PVDF composites to enhance dielectric properties, namely low dielectric constant as well as deprived interfacial compatibility [483,484]. Fu et al. reported the synthesis of various grain-sized BaTiO₃ particles via molten-salt synthetic route [485]. The AR-grade BaCO₃ (99.0%) and TiO₂ (99.5%) were utilized as initial materials. The NaCl–KCl (50 mol% NaCl + 50 mol% KCl) molten-salt medium was selected due its relative non-toxic nature and low eutectic melting temperature [486]. The detailed experimental procedure for the synthesis of BaTiO₃/polyvinylidene fluoride (BT/PVDF) nanocomposites was discussed and is shown in Figure 4.1. The PXRD profiles clearly exhibit pure BaTiO₃ phase at range of temperature between 650°C and 950°C. Further, the reduced eutectic point of the reaction containing molten salt, which owing to the low melting point of 670°C for NaCl–KCl.

As compared with conventional solid-state reaction, molten-salt route provides high reactivity as well as mobility of salts requires short reaction time and less synthesis temperature due to shortest diffusion distances of reaction mixtures. However, PXRD results of the sample calcined at more than 1,000°C shows small BaTi₂O₅ impurity peaks, which mainly attributed to increased volatility of molten salt [487]. In addition, phase tuning was clearly identified by fine scanning in the diffraction angle range $2\theta = 44.5°–46°$ (Figure 4.2a). The obtained result clearly exhibits phase

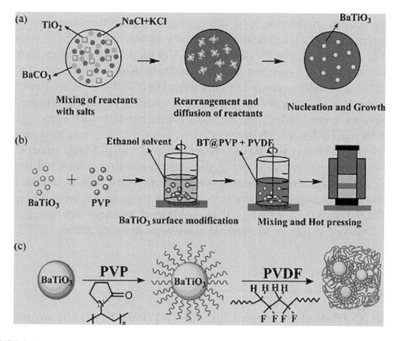

FIGURE 4.1
Schematic demonstration of synthesis, fabrication process, as well as surface modification mechanism of core-shell-structured PVP/BT-PVDF nanocomposites. (Reprinted with the permission from Ref. [485]. Copyright © ACS publications.)

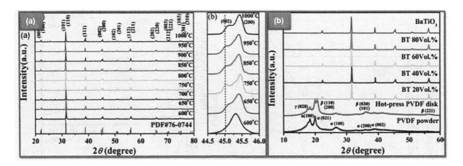

FIGURE 4.2
PXRD patterns of (a) BaTiO$_3$ particles synthesized at 600°C–1,000°C by molten-salt method, fine scanning XRD patterns of 2θ=44.5°–46° and (b) BaTiO$_3$ nanoparticles, pure PVDF powder, hot-press PVDF disk, and the composites with different volume fractions of BaTiO$_3$. (Reprinted with the permission from Ref. [485]. Copyright © ACS publications.)

tuning from pseudocubicto the tetragonal structure when increasing calcination temperature. Normally, the dielectric properties of BaTiO$_3$ strongly depend on the domain configuration as well as crystallite size. The polarization P$_s$ and tetragonality (c/a) can be related by the following equation [488]: $P_s \approx (c/a)^{0.5}$, where polarization is proportional to tetragonality (c/a), which is strongly associated to the particle size. The Rietveld refinements of synthesized BaTiO$_3$ particles clearly exhibit that gradual increment of tetragonality (c/a) with the elevated calcination temperature. Generally, ferroelectric properties of perovskite-type materials are mainly generated from the atomic displacement along ferroelectric axis as well as enhanced tetragonality (c/a). The calculated spontaneous polarization (P$_s$) value is found to be ~17 μC/cm^2 at 850°C, which increases with increase of calcination temperature. The highest P$_s$ value of 30 μC/cm^2 is noticed at 600 nm (950°C), indicating that the spontaneous polarization was strongly dependent on crystallite size [478]. In addition, XRD profiles of BaTiO$_3$, pure PVDF powder, hot-press PVDF disk, and PVP/BT-PVDF composites with different composition of BaTiO$_3$ were studied in detail. The XRD profiles of PVDF before and after hot-press exhibit characteristic hump of amorphous structure as compared to BaTiO$_3$ peaks (Figure 4.2b). In addition, transformation of α-phase into β-phase after hot-press process of PVDF powder was clearly revealed [489]. However, characteristic amorphous peaks belong to PVDF powder suppressed in nanocomposites profiles, which is mainly owing to shielding effect of intensive BaTiO$_3$ diffraction profiles.

The plot of dielectric permittivity and loss tangent as a function of BaTiO$_3$ volume fraction (Figure 4.3a) shows sudden increment in the dielectric constant with the increase of BaTiO$_3$ content. However, the same BaTiO$_3$ content in composites exhibit low dielectric loss, which indicates the less interfacial defects as well as homogeneous dispersion of PVP/BT particles in the PVDF matrix. When the volume of the BaTiO$_3$ powder increases to 80 vol%,

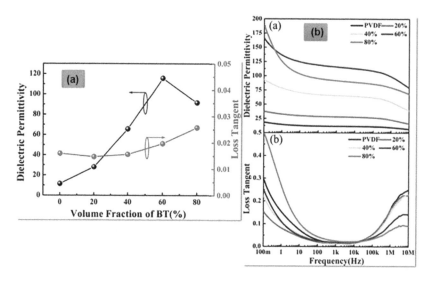

FIGURE 4.3
Dielectric permittivity and loss tangent of prepared samples as a function of volume fraction of BaTiO₃ and frequency. (Reprinted with the permission from Ref. [485]. Copyright © ACS publications.)

drastic decrement in the dielectric constant was clearly noticed. In addition, frequency reliance dielectric constant and loss tangent of pure PVDF polymer and the composites with different BaTiO₃ volume in the range 0.1 Hz to 10 MHz were systematically studied (Figure 4.3b). Interestingly, stable dielectric constant was noticed in all the samples between frequency ranges 1 Hz and 100 kHz. However, distinctive decrement of dielectric constant in the above 100 kHz range for the composites loaded with greater than 40 vol% of BaTiO₃ powder was revealed. This decrement in the dielectric constant in higher frequency range was mainly attributed to reduction in Maxwell–Wagner–Sillars (MWS) polarization as well as space charge polarization, which are more prominent in higher BaTiO₃ volume fraction composites [490,491]. However, high dielectric loss in the frequency range from 0.1 to 100 Hz was clearly noticed, which was due to the interfacial polarization.

The excellent dielectric performance of PVP/BT-PVDF composites was in agreement with previously reported works by tuning various parameters that greatly affect the dielectric properties of the composites [492,493]. In addition, various composite materials show that dielectric properties reported in previous literature was systematically listed in Table 4.1.

The prepared nanocomposite exhibits excellent spontaneous polarization that generated from the coarse size effect as well as enhanced interfacial compatibility owing to the typical core-shell structure, which results in high dielectric constant of the composites.

TABLE 4.1

Information of Particle Size, Surface Modification, and Dielectric Properties of Various Nanocomposites

Sl. No.	Composites	Particle Size (nm)	Surface Modification	Dielectric Permittivity	Dielectric Loss	Reference
1	$BaTiO_3$/ PVDF	100	-	50	0.03	Niu et al. [494]
2	$BaTiO_3$/ PVDF	-	N,N-dimethy lformamide	80	0.03	Zhou et al. [484]
3	$BaTiO_3$/ P(VDF-HFP)	100	PHFDA	37	0.02	Yang et al. [495]
4	$BaTiO_3$/ PVDF	100	NXT-105	42	0.04	Yu et al. [496]
5	$BaTiO_3$/ PVDF	100	-	45	0.03	Yu et al. [497]
6	$Pb(Zr,Ti)O_3$/ PVDF	160	-	50	0.025	Tang et al. [498]
7	$BaTiO_3$/ PVDF	100	PVP	50	0.09	Yu et al. [492]
8	$CaCu_3Ti_4O_{12}$/ Polyimide	-		50	0.01	Dang et al. [499]
9	$BaTiO_3$/ P(VDF-HFP)	100	Hydantoin resin	32	0.06	Luo et al. [493]
10	$BaTiO_3$/ PVDF	100	-	41	0.04	Dang et al. [500]

4.4 $BaTi_{1-x}M_xO_3$ (M=Cr, Mn, Fe, and Co) Nanocrystals

From past decades, many significant efforts have been paid to study the ferromagnetic properties of transition metal-doped $BaTiO_3$ structures [501–506]. As a generic example, Lin et al. and Venkata Ramana et al. [507] studied the ferromagnetic property of transition metal-doped $BaTiO_3$ structures in terms of the bound magnetic polaron model proposed by Coey and co-workers [508]. They found that the oxygen vacancies significantly induced magnetism in the materials by long-range interaction of the magnetic ions. Recently, Shuai et al. found that the Mn^{2+}-doped $BaTiO_3$ thin film show ferromagnetic property at room temperature with a greater oxygen vacancy concentration [509]. Mangalam and co-workers suggested that nonmagnetic ferroelectric $BaTiO_3$ show ferromagnetic behavior confirmed from density functional theory calculations as compared with their bulk counterparts [510]. The observed unexpected ferromagnetic behavior was owing to the existence of surface oxygen vacancies in the nanoparticles. As a result, two electrons will have pump Ti-electronic states from bottom of the conduction band of the nanomaterials thereby resulting in a spin polarization as well as a ferromagnetic coupling. Wang et al. reported analogous response in $PbTiO_3$ nanoparticles prepared via sol–gel method [511]. They experimentally found that the increment in the magnetism property of nanoparticles when their crystallite size decreased led to larger oxygen vacancies grafted on the surface of the nanoparticles. From the mentioned literature, it was clear that the surface of the particles plays a significant role in prompting ferromagnetic properties in nanomaterials. In 2018, Costanzo and co-workers synthesized transition metal-doped $BaTi_{1-x}M_xO_3$(M=Cr, Mn, Fe, and Co) cuboidal nanocrystals via solvothermal approach [512]. A procedure to synthesize $BaTiO_3$ colloidal nanocrystals doped with a particular transition metal ion is systematically given in detail in Figure 4.4a.

They found that morphology of the prepared nanocrystals is conserved irrespective of the doped transition metal ion (Figure 4.4b). Adjacently, morphological tuning of the prepared nanocrystals exhibiting cuboidal shape with a narrow size distribution via varying many reactions parameters,

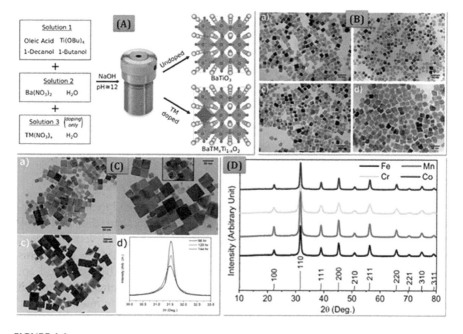

FIGURE 4.4
(a) Schematic illustration of synthesis procedure of prepared $BaTiO_3$:TM (TM=Cr, Mn, Fe, Co) colloidal nanocrystals, (b) TEM micrographs of TM-doped $BaTiO_3$ (TM=Cr, Mn, Fe, Co (4%)) colloidal nanocrystals, (c) TEM micrographs of $BaFe_{0.01}Ti_{0.99}O_3$ nanocrystals synthesized with various reaction time (96, 120, and 144 h) and plot of the most intense peak of the XRD patterns obtained from different reaction times, and (d) XRD profiles of TM-doped $BaTiO_3$ (TM=Cr, Mn, Fe, Co). (Reprinted with the permission from Ref. [512]. Copyright © ACS publications.)

namely reaction time as well as polarity of the solvent, was studied in detail (Figure 4.4c). The crystallite size of the prepared nanocrystals was found to be ~10–70 nm by altering the reaction duration. The XRD profiles of doped BaTiO$_3$ nanocrystals match well with the standard XRD pattern of cubic phase (PDF# 31-074), indicating that the transition metal ions were replaced in the titanium ion sites (Figure 4.4d) [513]. The dielectric data of the prepared BaTiO$_3$:Fe nanocrystals reveal the decrement of dielectric with Fe concentration in the samples (Figure 4.5a). Remarkably, a broad peak belonging to Fe-doped and undoped sample in the temperature range from 80°C to 100°C was clearly observed, which shows the ferroelectric to para-electric phase transition (Figure 4.5c). The obtained experimental results clearly reveal distorted intrinsic tetragonal structure of a sample at room temperature, which results in the creation of electrical dipoles coupled to the ferroelectric property of BaTiO$_3$ nanocrystals. Further, the highest peak was found in a lower temperature, which indicates the decrement of Curie temperature upon Fe doping in BaTiO$_3$ nanocrystals (Figure 4.5). The ferroelectric behavior of the prepared samples was extensively studied by employing DART-PFM method [514], in which tip as a top electrode on nanocrystals thin film drop-casted on conductive silicon (Figure 4.5). The DART-switching spectroscopy mode (DART SS-PFM) data of the prepared sample show hysteretic nature in the phase signal (phase shift of 180°), as anticipated from the induced dielectric polarization [515]. The amplitude signal exhibit the characteristic ferroelectric property of a material with a butterfly-like shape (Figure 4.5d), which further confirms the ferroelectric nature of the sample. Furthermore, the ferroelectric behavior of the prepared crystals also examined by writing two-square pattern with the PFM tip in contact mode while applying a bias (–10 and +10 V) at the tip. Both the obtained both clear square patterns clearly indicate the existence of both upward- as well as downward-oriented polarization, which can be reversible by applying electrical field. The topographic image completely uncorrelated as compared with the PFM phase, which endorses that the witnessed phase shift is not affected by cross-talk effects and is the actual image of the sample polarization. Furthermore, VSM data confirmed that both Fe-doped as well as pristine samples are ferromagnetic and the saturation magnetization (M$_s$) increases with Fe doping, which are in good agreement with previously literature [516–518]. Particularly, the value of Ms of the BaTi$_{0.94}$Fe$_{0.06}$O$_3$ nanocrystal was found to be 1×10^{-4} emu/g, whereas values of 1.5×10^{-4} and 2.5×10^{-4} emu/g were obtained for BaTiO$_3$:Fe (2 and 4%). From experimental VSM and EPR data comparison, presence of multiple coupling mechanisms was confirmed. The obtained aforementioned results confirm that BaTiO$_3$:Fe nanocrystals show both ferromagnetism as well as ferroelectricity at room temperature, which makes them the probable candidate for multiferroics materials.

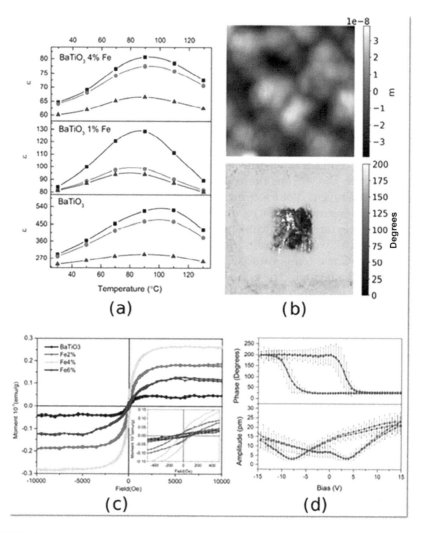

FIGURE 4.5
Ferroelectric and ferromagnetic properties of Fe-doped BaTiO$_3$. (Reprinted with the permission from Ref. [512]. Copyright © ACS publications.)

4.5 BaTiO$_3$/Polydimethylsiloxane (PDMS) Nanogenerators

In recent years, progress of substitute viable energy exceptionally urgent need due to possible diminution of fossil energy in the near future. Although, naturally available materials, namely wind, solar, and hydro-power can significantly alleviate and are not available at any period and affected by many environmental factors [519–521]. To overcome such limitations, piezoelectric nanogenerators are one of the most prominent energy providers for microsystems by gather electrical charges in materials as a function of mechanical stress. The created energy is sufficient enough to power LEDs, biosensors, micro-electromechanical systems, mobile sensors, wearable personal electronics, as well as ultra-low power consumption wireless electronics systems [522–525]. Furthermore, the piezoelectric nanogenerators can also be coupled with capacitors or batteries, in which mechanical energy is directly converted and stored as chemical energy [526]. From the past decades, various geometrical structures, namely nanosheets, nanotubes, nanowires/fibers, nanoparticles, triangular belts, nanorods, and nanocubes, have been broadly investigated for piezoelectric energy-harvesting ceramics [443,527–529]. Among these, nanofibers offer enhanced mechanical robustness as well as obedience to less loads for efficiently converting mechanical energy from low-intensity strain [530]. In 2016, Yan and co-workers prepared $BaTiO_3$ nanofibers via electrospinning method using sol–gel precursor followed by calcination process [531]. Furthermore, three different types of flexible piezoelectric nanogenerators were fabricated by aligning $BaTiO_3$ nanofibers in PDMS elastomer matrix. The systematic synthesis procedure for preparing $BaTiO_3$ nanofibers and $BaTiO_3$/PDMS nanogenerators is shown in detail in Figure 4.6a and b. The obtained results show that the followed technique is comparatively very simple, economic, as well as scalable for industrial range for the fabrication of well-aligned and controlled nanofibers. The morphological results of final $BaTiO_3$ nanofibers show fine-aligned integrated structure as well as irregular fiber surface with average diameter of ~354.1 nm, which ascribed to the well crystallinity of $BaTiO_3$ and PVP degradation through calcination. On the other hand, cross-section view of SEM images of the $BaTiO_3$/PDMS composites confirms that nanofibers are vertically aligned in the PDMS matrix without any traces of voids or pores. The homogeneous dispersion of nanofibers in PDMS matrix was clearly revealed. Even though $BaTiO_3$ nanofibers are delicate and tough to handle after calcination, $BaTiO_3$/PDMS composites show excellent flexibility, which endorses the applicability of the composites as piezoelectric nanogenerators. The variation of dielectric constants of pure PDMS and BTNF-R as a function of frequency (Figure 4.7a) was found to be 3.85 and 4.14 at 100 Hz, respectively. However, the dielectric constants of BTNF-V and BTNF-H are estimated and obtained to be 40.23–35.88 and 23.99–20.72 in the frequency range from 100 Hz to 2 MHz, respectively.

The obtained result realizes that the $BaTiO_3$ nanofibers in BTNF-V and BTNF-H have greater aspect ratio attributed to their lowest percolation thresholds, as compared to BTNF-R with short $BaTiO_3$ nanofibers [532]. Interestingly, the dielectric constants of BTNF-V as a function of applied frequency are nearly two-fold higher as compared to BTNF-H. The obtained

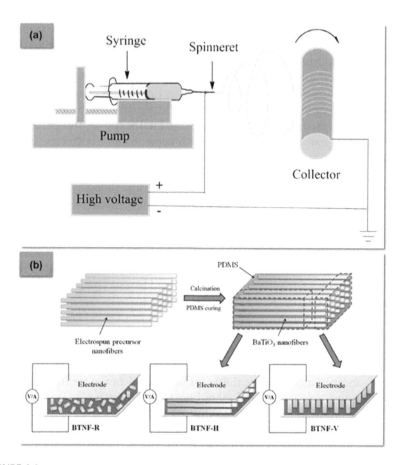

FIGURE 4.6
The schematic illustration of (a) preparation of uni-axially aligned BaTiO$_3$ nanofibers via electrospinning technique and (b) fabrication procedure BaTiO$_3$ nanofibers centered nanogenerators in three alignment modes with piezoelectric test circuits. (Reprinted with the permission from Ref. [531]. Copyright © ACS publications.)

result clearly indicated that the BTNF-V possesses higher density of BaTiO$_3$ than BTNF-H along the test direction. In conclusion, the orientation of BaTiO$_3$ nanofibers in the PDMS matrix greatly affects the dielectric permittivity as the influence of electric field is greater along the direction of BaTiO$_3$ fiber. However, dielectric loss tangents of BTNF-R, BTNF-H, and BTNF-V are found to be 0.006, 0.062, and 0.042 at 100 Hz, respectively. It was well-known that the loss tangent is the proportion of the transferred charges by conduction to that stored by polarization [533].

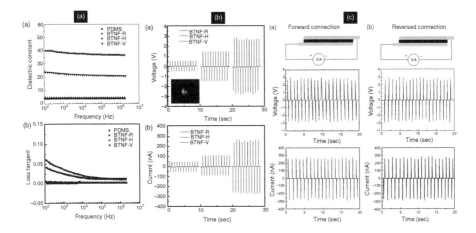

FIGURE 4.7

(a) Dielectric constant and loss tangent of prepared samples as a function of frequency, (b) Output voltage, current variations of BaTiO$_3$/PDMS nanogenerators under a periodic mechanical compression. Inset: photograph of a commercial blue LED lit up by the electric energy generated from BTNF-V, and (c) Output voltage and current changes of BaTiO$_3$/PDMS nanogenerators in forward direction and reverse direction under a periodic mechanical compression. (Reprinted with the permission from Ref. [531]. Copyright © ACS publications.)

The detailed investigations of piezoelectric behavior of prepared samples were performed by applying a mechanical pressure of 0.002 MPa (Figure 4.7b). The output voltage of the three differently aligned BTNF-R, BTNF-H, and BTNF-V nanogenerators are found to be ~0.56, 1.48 and 2.67 V, respectively. This attributed to the effects of the threshold percolation as well as the alignment direction between electrodes. However, the average output current of BTNF-R, BTNF-H, and BTNF-V are obtained to be ~57.78, 103.33, and 261.40 nA, respectively, which showed similar variation profiles than that of the output voltage. Furthermore, output power produced from the BTNF-V system was highly sufficient to commercial blue LED emission was noticed (Figure 4.7b). The piezoelectric behavior of BTNF-V in the forward as well as reversed directions was systematically discussed (Figure 4.7c). The results clearly indicate that the output voltage and current of the fabricated nanogenerators exhibit opposite values, which endorse that the obtained voltage and current signals are produced by the piezoelectric property of nanogenerators, not by other parameters, such as noise or environment effect. The overall results clearly endorse that the BaTiO$_3$ nanofiber-centered flexible nanogenerators have potential applications in various fields such as sensors and storage elements for wireless as well as self-powered and mechanical energy-harvesting devices. The fabrication, structural characterization, and energy-harvesting properties of BaTiO$_3$-centered piezoelectric nanogenerators are tabulated in Table 4.2.

TABLE 4.2

Details of Energy-Harvesting Properties of $BaTiO_3$ Perovskites-Centered Piezoelectric Nanogenerators

Sl. No	Sample	Morphology	Synthesis Method	Working Mode	Output Voltage (V)	Output Current (nA)	References
1	Single $BaTiO_3$ nanowire	Nanowire	Electrospinning	Bending amplitude of 15 mm	0.21	1.3	Ni et al. [534]
2	Vertically aligned $BaTiO_3$ nanowire arrays	Nanowire	Hydrothermal	Vibration excitation of 1 g acceleration	0.312	0.9	Koka et al. [443]
3	Single $BaTiO_3$ nanowire/ PVC fiber	Nanowire	Topochemical	Bending on figure	0.9	10.5	Zhang et al. [535]
4	$BaTiO_3$ thin film	-	-	Stress of 340 MPa	1	26	Park et al. [521]
5	Highly oriented $BaTiO_3$ film	Micro-platelet	Topochemical microcrystal conversion	Tapped by a paddle	2.3	25	Gao et al. [536]
6	$BaTiO_3$/ MWCNT/ PDMS generator	Nanoparticle	Hydrothermal	Stress of 0.057 MPa, Strain of 0.33%	1.5 3.2	150 250-350	Park et al. [537]
7	$BaTiO_3$/PDMS Nanogenerator	Nanotube	Hydrothermal	Stress of 0.2 MPa Stress of 0.4 MPa Stress of 1.0 MPa	1.0 3.1 5.5	- - 350	Lin et al. [538]
8	Virus-templated $BaTiO_3$ nanogenerator	-	Genetically programmed self-assembly	Curvature radius: 10.5 cm Curvature radius: 7.5 cm Curvature radius: 6 cm Curvature radius: 5 cm	2 3 5 6	- - - 300	Jeong et al. [539]
9	$BaTiO_3$/P(VDF-HFP) thin film	Nanoparticle	-	Stress of 0.23 MPa Bending	75 5	1.5×10^4 600	Shin et al. [540]

4.6 CaTiO$_3$:Eu^{3+} Phosphors

The luminescence properties of the ATiO$_3$ (A = Pb, Ca, Sr, and Ba) perovskite structure has been paid much attention due to their special thermal, optical, mechanical, as well as electrical properties, which can be utilized to fabricate most significant phosphors [328,393,541,542]. The idiosyncratic perovskite structural deformation normally depresses the point symmetry at the alkaline-earth site in (Ca, Sr, Ba)TiO$_3$:RE^{3+} [543–545]. This leads to enhancement in f–f transition of RE^{3+} which makes the titanate host suitable for solid-state display technologies. The impurities cause photoluminescence (PL) in the host material, which serve as an activator, also called PL center. The fraction of doped impurities in the material activate the extrinsic PL behavior. Transition metal ions with an electron configuration of 3d can interact strongly with the crystal field. This results in changes in the energy level structure of the free ion. Generally, trivalent and divalent RE ions act as activators. These RE ions are responsible for characteristic narrow PL emission bands, which were attributed to f*→f transitions. The narrow PL emission band of the RE ions originates in the 4f sub shell, which is partially shielded by [5s^{2}5p^6] electrons [546,547]. Since the interaction between the 4f electron and the crystal field is weaker than the spin-orbital interaction, the energy levels of the RE ions with a 4f$^{(k-1)}$4f^{1*}configuration are not significantly influenced by the crystal fields. Therefore, for RE ions with a 4f$^{(k-1)}$4f^{1*}configuration, the structure of the energy levels in the free RE ions is basically the same in different host matrices. Som et al. studied the preparation of Eu^{3+}-doped and Eu^{3+}/K$^+$ co-doped CaTiO$_3$ phosphors via solid-state reaction method [548]. The orthorhombic distorted perovskite structure with the space group Pb was confirmed from XRD profiles. The lattice parameters of the unit cell are elucidated and found to be a = 5.380 Å, b = 5.440 Å, and c = 7.639 Å which are in accordance with the standard JCPDS no. 86–1393. Furthermore, a slight peak shift (i.e., (112) plane) upon increase of the Eu^{3+} concentration was clearly noticed, which specifies that the Eu^{3+} doping does not affect the host CaTiO$_3$ lattice [549]. The charge compensation in any phosphor greatly affects the optical activity. The charge compensation in CaTiO$_3$:Eu^{3+}arises by the generation of negatively charged Ca and positive oxygen vacancies. In orthorhombic lattice of the CaTiO$_3$:Eu^{3+} phosphor, Ca^{2+} and Ti^{4+} ions are coordinated with eight and six O atoms, respectively, which results in distorted dodecahedron and octahedron structures [550]. The substitution of Eu^{3+} ions in the dodecahedron sites (CaO$_8$) of Ca^{2+}, whereas Ti^{4+} remains unaltered in the octahedral sites (TiO$_6$), was clearly illustrated. In the case of charge compensation, 3Ca^{2+} has been substituted by 2 Eu^{3+}, which results in a Ca^{2+} vacancy:

$$3Ca^{2+} \rightarrow 2Eu^{3+} + VCa^{2+}$$

Furthermore, charge compensation by K^+ ion results in the following equation:

$$2Ca^{2+} \rightarrow Eu^{3+} + K^+$$

The PL excitation spectrum of $CaTiO_3$:Eu^{3+}phosphors upon emission at 617 nm shows a broad charge-transfer (CT) band at ~235 nm, which ascribed to a transfer of electron from the oxygen 2p orbital to the empty 4f orbital of Eu^{3+} [551,552]. Later, various sharp peaks at ~360, 380, 397, 417, 465, and 535 nm, owing to the $^7F_0 \rightarrow$ $^5D_4$, $^5L_7$, $^5L_6$, $^5D_3$, $^5D_2$, and $^5D_1$ transitions of Eu^{3+} ions, respectively, were revealed. Further, absence of CT in present material was clearly noticed as compared with other traditional Eu^{3+}-doped samples. This was quite unusual, since the typical Eu^{3+}-doped samples always exhibit strong CT transition absorption band around ~200–300 nm. The results clearly evidenced that the present system can be readily excited by near-UV light as well as blue light, which signifies the potentiality of phosphor in UV LED-converted solid-state lighting technology [553]. However, PL emission spectra show major emission peaks at 514, 540, 595, and 617 nm, which attributed to $^5D_1 \rightarrow$ $^7F_3$ and $^5D_1 \rightarrow$ $^7F_2$, $^5D_1 \rightarrow$ $^7F_1$ and $^5D_1 \rightarrow$ $^7F_0$ transitions of Eu^{3+} ions, respectively. Among peak at ~617 nm ascribed to electric–dipole transition, which hypersensitive to symmetry of the host structure. Furthermore, peak at ~595 nm belong to magnetic dipole transition and is insensitive to site symmetry [554]. Asymmetric ratio is one of the important parameters in order to measure the site symmetry of Eu^{3+} in $CaTiO_3$ host. The luminescent behavior of the phosphors is highly influenced by the disparity of the asymmetric ratio, which can be altered with doping concentration. Further, determining the optimum concentration of the RE^{3+} ions in a host for luminescence application was much essential. The variation of PL intensity as a function of Eu^{3+} concentration exhibits increased emission intensity up to 3 mol% and later diminishes. The rapid decrement in PL emission is ascribed to concentration quenching in the host. Further, the quenching phenomena can be explained by a modification of the energy levels in the host material, which makes the PL activation very inefficient. At higher activator concentrations, the PL property diminished by the dopant itself attributed to resonant absorption processes. There are three important types of energy transfer mechanisms which are responsible for concentration quenching; these are discussed in detail.

   a. Electron–phonon coupling and multi-phonon (MP)-assisted energy transfer
     The distance between two $^5D_2$ and $^5D_3$ energy levels at ~2500 cm^{-1} is larger than vibrational energies of chemical bonds in many inorganic materials [555]. Furthermore, weak bands are situated in between 595 and 617 nm related to the phonon-assisted transition (PAT) [556]. Normally, one or MP processes can be responsible in PAT. In the present work, the cut-off phonon energy of $CaTiO_3$ is ~550 cm^{-1}.

The Huang–Rhys factor of the prepared samples was calculated and found to be ~0.18. This value evidences the presence of weak electron–phonon coupling between Eu^{3+} and $CaTiO_3$ host. Therefore, MP-aided energy transfer is not the basis of quenching.

b. *Eu^{3+} to Ca^{2+} vacancy energy transfer*
Based on the vacancy model, the increase of the Eu^{3+} concentration diminishes the PL emission in this host by the formation of electron-captured Ca vacancies, as described in the following equations. The substitution of Eu^{3+} ion in Ca^{2+} site upon Eu^{3+} is doped in host matrix as illustrated below [548];

$$2\left[EuO_{12}\right]'' + \left[V_{Ca}''O_{12}\right]_{complex} \rightarrow 2\left[EuO_{12}\right]^{x} + \left[V_{Ca}^{x}O_{12}\right]_{complex}$$

where $\left[EuO_{12}\right]''$ and $\left[V_{Ca}''O_{12}\right]_{complex}$ are donor and accepter, respectively. This equation corresponds to the neutrality of the lattice charge for the unexcited phosphors. The generation of self-trapping of electrons is probable upon excitation by electromagnetic radiation (~397 nm) in the following way [548];

$$\left[V_{Ca}^{X}O_{12}\right]_{Complex} + e_{Excited} \rightarrow \left[V_{Ca}'O_{12}\right]_{Complex}$$

$$\left[V_{Ca}'O_{12}\right]_{Complex} + e_{Excited} \rightarrow \left[V_{Ca}''O_{12}\right]_{Complex}$$

Hence, there is diminished PL intensity via energy transfer process between Eu^{3+} ions and Ca^{2+} vacancies [557]. This type of energy transfer was systematically studied.

c. *Eu^{3+} to Eu^{3+} energy transfer and concentration quenching*
The Eu^{3+}–Eu^{3+} ions energy transfer can generally arise owing to multipole–multipole interaction, radiation re-absorption, as well as exchange interaction [558,559]. The type of non-radiative energy transfer involved in concentration quenching was elucidated by critical distance (R_c) between the two neighboring Eu^{3+} ions by using Blasse's relation [560]. The R_c value of the present system was estimated and found to be 24.2 Å. The obtained value of R_c is greater than 5 Å, suggesting that the multipolar interaction is responsible for concentration quenching of Eu^{3+} ions in the $CaTiO_3$ host. There are various types of multipolar interactions, namely dipole–dipole (d–d), dipole–quadrupole (d–q), quadrupole–quadrupole (q–q), etc. The know the authorative interaction responsible for energy transfer was estimated using the Dexter and Schulman relation [561]. The calculated value of s (series of the electric multipolar) was found to be 5.94,

which is nearly equal to 6 (for d–d interaction). The obtained result revealed that the d–d interaction is mainly liable for the concentration quenching of Eu^{3+} ions in $CaTiO_3$ phosphors. The enhancement in PL emission upon co-doping of charge compensator (i.e., K^+) in the host was clearly noticed. The photometric characterization (CIE, CCT, and CP) indicated the suitability of $CaTiO_3$: Eu^{3+}, K^+ phosphor for pure red emission in light-emitting diode applications.

4.7 Zn²⁺ Ions in BaTiO₃: Er³⁺/Yb³⁺ Nanophosphor

The materials doped with RE ions create much interest in research community due to its significant applications in various fields, such as fiber-optic communication, display devices, solid-state lighting, and therapeutic as well as biomedical applications [562–564]. In recent years, the fabrication of economic and consistent multifunctional materials motivated researchers [565]. Normally, RE ions exhibit both down-conversion (Stokes type) and up-conversion (anti-Stokes type) luminescence and the advantage of both types is already well demonstrated [566]. Among those, anti-Stokes-type luminescence is considered to be more strong in low phonon frequency hosts, especially oxides are superior for doping of lanthanides due to their excellent durability as well as chemical stability. $BaTiO_3$ perovskite is a most favorable ferroelectric material, which offers excellent functionality toward phosphor-based wide range of applications. The transition metal ion as dopants in phosphors in altering the properties of $BaTiO_3$ is found to be of more technological interest. The Zn^{2+} ions favor localization in the $BaTiO_3$ lattice, owing to small ionic radius [567]. By considering these, Mahata et al. synthesized Zn^{2+} in $BaTiO_3$:Er^{3+}/Yb^{3+} nanophosphor ions via wet-chemical co-precipitation method by maintain the composition of the nanophosphor as given below [568]:

$$(100 - x - y - z)\,mol\%\,BaTiO_3 + x\,mol\%\,Er_2O_3 + y\,mol\%\,Yb_2O_3 + z\,mol\%\,ZnO$$

where $x=0.3$, $y=3.0$, and $z=0, 10, 20$. The detailed synthesis procedures of the prepared samples were discussed in detail. The cubic phased-$BaTiO_3$ was confirmed from PXRD results and no additional peak is observed. The peak shift toward higher angle side was clearly evidenced upon Zn-doping (20 mol%), which shows unit cell expansion owing to ionic size mismatch of Zn^{2+} and Ti^{4+} ions. On the basis of thermodynamic and tolerance factors, the typical $BaTiO_3$ perovskite structure with cubic packing of Ba^{2+}, O^{2-} and Ti^{4+} ions filling the octahedral holes of the crystal was discussed in detail [569]. The diminished optical band gap particularly from 3.61 to 3.35 eV upon Zn^{2+} doping was noticed. This nature was successfully explained based on the average bond energy of the material [570]. The UC PL emission spectra of prepared phosphors upon excitement by 980 and 800 nm at two different temperatures (280 and 12 K) show emission peaks at 411, 524, 550, and 661 nm. The 2- and 5-times enhancement of green and red emission peaks in Zn^{2+}-doped samples was clearly noticed upon 980 nm excitation wavelength. However, enhancement in intensity of green and red emission around 12 and 2 times were observed under 800 nm excitation at 12 K temperature. Furthermore, increment in asymmetric ratio between red to green bands by introduction of Zn^{2+} ions after 980 nm excitation was revealed, which specifies the cross-relaxation population of the $^4I_{13/2}$ level of the Er^{3+} ion [571]. The observed intensity enhancement by Zn^{2+} doping was mainly due to alterations of crystal field symmetry environment around the RE^{3+} ions. The UC emission mechanisms by considering various processes [344,572] was explained in detail. Aforementioned results exhibit affordable potential multifunctional applications of the samples.

4.8 $MgTiO_3$:Mn^{4+}

In 2018, Glais et al. synthesized Mn^{4+}-doped $MgTiO_3$ perovskite nano powders by using molten salt assisted sol–gel route [573]. The XRD profiles of prepared $MgTiO_3$:Mn^{4+} nanoparticles show rhombohedral crystal structure (Figure 4.8a). From TEM images, excellent dispersion of square

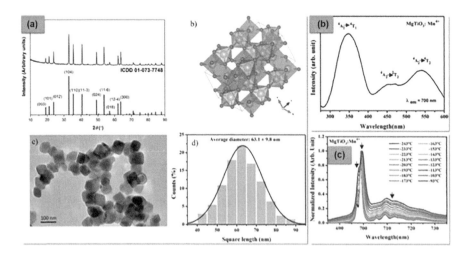

FIGURE 4.8
(a) PXRD, schematics of the $MgTiO_3$ crystal structure, TEM, and its associated distribution diagram, (b) Excitation spectrum, and (c) Emission spectra of $MgTiO_3$:Mn^{4+} at various temperatures. (Reprinted with the permission from Ref. [573]. Copyright © RSC publications.)

nanoparticles with average square length is 63.1 ± 9.8 nm was clearly observed. The octahedral positions where usually Ti^{4+} ions situated in the rhombohedral crystal structure can be easily substituted by dopant Mn^{4+} ions due to high stabilized ligand field energy in octahedral coordinated systems (Figure 4.8a). The excitation spectrum of $MgTiO_3$:Mn^{4+} nanoparticles show broad bands at ~350, 455, and 550 nm upon emission wavelength of 700 nm (Figure 4.8b). The obtained bands were attributed to $^4A_2 \rightarrow {}^4T_1$, $^4A_2 \rightarrow {}^4T_2$, and $^4A_2 \rightarrow {}^4T_2$ transitions.

Further, emission spectra of the prepared samples are recorded under 550 nm excitation between –250°C and 50°C exhibit intensive red emission centered at ~700 nm (Figure 4.8c), which attributed to spin forbidden $^2E \rightarrow {}^4A_2$ transition. Two emission intensity ratios can be used to get precise temperature measurements on two different temperature ranges. The emission intensity ratio in temperature ranges from –250°C and –90°C exhibit a maximum sensitivity and thermal resolution of 1% °C^{-1} and 0.42°C. However, intensity ratios between $^2E \rightarrow {}^4A_2 / {}^4T_2 \rightarrow {}^4A_2$ transitions in the temperature range of –200°C to 50°C display maximum sensitivity of 0.6% °C^{-1}. As compared with other luminescence-based temperature measurement probes, the present system exhibits higher temperature resolution value of 0.025°C (at 4°C) (Table 4.3).

According to obtained results, the prepared $MgTiO_3$:Mn^{4+} nanoparticles are considered as better material for multi-range luminescence temperature probes.

TABLE 4.3

Comparison of Temperature Resolutions (δT) of Some Inorganic Luminescence Temperature Nanoprobes

Material	Sensitivity (% °C^{-1}) (temperature (°C))	δT (°C)	References
TiO_2:Eu^{3+}	2.43 (250)	0.46	Nikolić et al. [574]
$GdVO_4$@SiO_2:Tm^{3+}, Yb^{3+}	0.94 (50)	0.40	Savchuk et al. [575]
Gd_2O_3:Nd^{3+}	1.75 (15)	0.10	Balabhadra et al. [576]
CaF_2:Gd^{3+}, Nd^{3+}	0.12 (25)	1.80	Cortelletti et al. [577]
$Bi_2Ga_4O_9$:Cr^{3+}	0.7 (37)	0.08	Back et al. [578]
$LiLaP_4O_{12}$:Cr^{3+}, Nd^{3+}	4.89 (50)	-	Marciniak et al. [579]
$MgTiO_3$:Mn^{4+}	4.1 (4)	0.025	Glais et al. [573]
$MgTiO_3$:Mn^{4+}	0.6 (–125) 1.2 (–180)	0.42 (at –180°C) 0.84 (at 50°C)	Glais et al. [573]

4.9 CaTiO₃:Pr³⁺@SiO₂

Li and co-workers synthesized long-persistent CaTiO₃:Pr³⁺ red-emitting material via solid-state reaction route [580]. The prepared CaTiO₃:Pr³⁺ material was functionalized by coating silica shell to improve the solubility of the material in several solvents. The detailed synthesis procedure for the preparation of the CaTiO₃:Pr³⁺ and CaTiO₃:Pr³⁺@SiO₂ nanoparticles was systematically presented. The morphological studies of CaTiO₃:Pr³⁺@ SiO₂ nanoparticles display uniform dispersion with diameters of 115–300 nm and the average size was 201 nm (Figure 4.9a). The XRD profiles of CaTiO₃:Pr³⁺ nanoparticles were indexed to (112), (220), (312), (224), and (332) planes, which were well matched with JCPDS card no. 01–081–0561 of perovskite. However, XRD profile of CaTiO₃:Pr³⁺@SiO₂ nanoparticles display an abroad peak at diffraction angle ~ 21°, which belongs to the

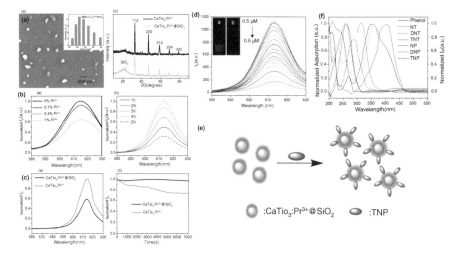

FIGURE 4.9
(a) PXRD and SEM images, (b) Phosphorescence emission spectra of CaTiO₃:Pr³⁺ (0%, 0.1%, 0.4% and 1%) nanoparticles and CaTiO₃:Pr³⁺ (0.4%) nanoparticles with different calcination temperature, (c) Phosphorescence emission spectra of CaTiO₃:Pr³⁺@SiO₂ nanoparticles and the photo-stability, (d) Phosphorescence spectra of CaTiO₃:Pr³⁺@SiO₂ upon addition of increasing concentrations of TNP (0.5–300 μM), (e) Scheme for the phosphorescence quenching for detection of TNP, and (f) Spectral overlap between excitation spectrum of CaTiO₃:Pr³⁺@SiO₂ and absorption spectra of different nitro-explosive compounds in PBS solution. (Reprinted with the permission from Ref. [580] Copyright © RSC publications.)

amorphous SiO_2 shells (Figure 4.9a). The phosphorescence emission spectra of $CaTiO_3$:Pr^{3+} (0, 0.1, 0.4 and 1%) nanoparticles under 315 nm excitation wavelength display intensive red emission at 614 nm, which can be attributed to $^1D_2$ $^3H_4$ transition of Pr^{3+} ions in 0.4% doped sample (Figure 4.9b). Similarly, phosphorescence emission spectra of $CaTiO_3$:Pr^{3+} (0.4%) nanoparticles calcined at 900°C for 3 h presented the highest phosphorescence intensity. However, silica-coated $CaTiO_3$:Pr^{3+} (0.4%) nanoparticles show almost 30% diminished phosphorescence intensity within 2 h (Figure 4.9c). The obtained results concluded that the core-shell nanoparticles had superior photo-stability than $CaTiO_3$:Pr^{3+} (0.4%) nanoparticles. Furthermore, the gradual decrease of phosphorescence intensity of $CaTiO_3$:Pr^{3+}@SiO_2 nanoparticles with increasing 2,4,6-trinitrophenol (TNP) concentrations was clearly observed. The obtained result presented a good linear nature between phosphorescence and TNP concentrations ranging from 0.5 to 100 mM and the detection limit was found to be 20.6 nM (Figure 4.9d). The obtained PL property was mainly due to electromagnetic interactions as well as inner filter effect. The probable mechanism involved in detection of TNP by phosphorescence quenching based on the $CaTiO_3$:Pr^{3+}@SiO_2 nanoparticles was discussed in detail (Figure 4.9e). In addition, the present system showed excellent selectivity toward TNP as compared with similar structured compounds, namely 4-nitrotoluene (NT), 4-nitrophenol (NP), 2,4,6-trinitrotoluene (TNT), 2,4-dinitrophenol (DNP), phenol, and 2,4- dinitrotoluene (DNT) (Figure 4.9f). The prepared sample tested effectively to detect the TNP concentrations in water samples and displayed its potential application for environmental pollution analysis.

5

Applications of Titanate Materials

5.1 BaTiO₃: RE Yellow Phosphors an TL Applications

Barium titanate ($BaTiO_3$) is a ferroelectric oxide, and it is crystal clear in visible and infrared light and has strong electro-optic coefficients making it appealing for dynamic and aloof optical parts [581]. $BaTiO_3$ has been viewed as a charming host material for radiant dopants in that it is wide band gap energy and it is about 3.38 eV when the light's polarization is corresponding to the c-axis and it is about 3.27 eV when the polarization is opposite to the c-axis. It is an electro-optical material that has commonly high strong solubilities for a grouping of lanthanide activator cations, for instance Ce, Nd, Sm, Eu, Gd, Dy, Ho and Er [582–585]. $BaTiO_3$ nanoparticles have been synthesized by utilizing hydrothermal technique, sol–gel preparation, composite-hydroxide-intervened strategy, fire-helped splash pyrolysis (FHSP), radio frequency plasma chemical vapor deposition (RF-plasma chemical vapor deposition (CVD)), one-stage sol-precipitation route, and electrochemical route. Pure $BaTiO_3$ is an insulator though after doping it is anything but a semiconductor. Other than Positive Temperature Co-efficient of Resistivity (PTCR) properties, semiconductor $BaTiO_3$ is utilized in the sensor applications. $BaTiO_3$ is an exceptionally helpful host matrix for certain applications since it does not absorb energy, permitting it to provide only for sustenance of rare earth (RE) ions, which are very useful for these exacting applications. Especially, RE-doped near infrared (NIR)-to-visible ceramic oxides correspond to an unconventional and excellent substitute for conventional fluorescent applications [586].

RE ions are trivalent in character most of the compounds and in the current situation which dwell in Ba sites in $BaTiO_3$. Charge equilibrium in the system could be remunerated by electrons, A-site Ba vacancies, B-site Ti vacancies, or the $^3d_1$ (Ti^{3+}) mode, are owing to the different synthesis methods and ratios of starting chemical compositions [587]. $BaTiO_3$ is a precious host matrix for a few applications because it does not absorb energy, permitting it to provide merely for sustenance of lanthanides, which is valuable for

DOI: 10.1201/9781003381907-5

these meticulous applications. Particularly, RE-doped NIR-to-visible ceramic oxides symbolizes another and outstanding replacement for conventional fluorescent applications [588,589]. In the middle of the RE ions, trivalent dysprosium (Dy^{3+})-activated glasses have been disputably hopeful materials for two-color phosphors and white light emission since Dy^{3+} ion inherits intense emissions at blue (488 nm, $^4F_{9/2} \to {}^6H_{15/2}$), green (544 nm, $F_{9/2} \to {}^6H_{13/2}$), and red (621 nm, $^4F_{9/2} \to {}^6H_{11/2}$) regions. Further, it is well known that the $^4F_{9/2} \to {}^6H_{13/2}$ transitions of Dy^{3+} ions are hypersensitive and therefore its intensity strongly depends on the nature of the host, whereas the intensity of magnetic dipole-allowed $^4F_{9/2} \to {}^6H_{15/2}$ transition is less sensitive to the host. Hence, at a suitable environment, the intensity ratio of these green-to-blue (G/B) transitions will be modified resulting in the Dy^{3+}-activated phosphors producing white light [367,589].

5.1.1 X-Ray Diffraction (XRD) of $BaTiO_3$: Dy^{3+} Phosphor

The XRD pattern of $BaTiO_3$: Dy^{3+} phosphor prepared using hydrothermal synthesis technique is displayed in Figure 5.1. All the diffraction peaks of the sample at 2θ are in good agreement with Bragg diffraction were found at (011), (100), (002), (020), (111), (102), (120), (022), and (200) certifies the configuration of orthorhombic crystal structure of $BaTiO_3$ which corresponds to the crystallographic open database Card Number 96-901-4775. The fitted XRD pattern matched with the standard pattern of $BaTiO_3$ which is revealed in Figure 5.1. The crystallite size of the sample erstwhile computed from the

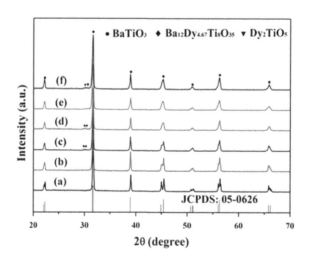

FIGURE 5.1
XRD pattern of Dy^{3+}-doped $BaTiO_3$ phosphor. (Reprinted with the permission from Ref. [594]. Copyright © Springer publications.)

full-width half-maximum (FWHM) of all peaks in XRD pattern using the Scherer formula is given by [593]:

$$D = \frac{0.9\lambda}{\beta Cos\theta D}$$

where D is the crystallite size at a 90° angle to the reflecting planes, $\lambda = 0.154$ nm correspond to the wavelength of X-ray source, θ is the diffraction angle, and β represents the FWHM of diffraction peak. The crystallite size for intense peak in the XRD pattern of synthesized by $BaTiO_3$: Dy^{3+} phosphor was found 41.81 nm.

5.1.2 TL Glow Curve Studies

Thermoluminescence (TL) glow curve investigation methods to determine their ability for applications of high γ-dose dosimeter for industrial and scientific areas. Exposure to γ-radiation doses from Co_{60} was assorted between 1 and 2.5 kGy. The glow curves acquired displayed emission peak temperatures deceitful between 100°C and 300°C and were establish to be dose dependent (Figure 5.2a). The emission peak temperature for apiece dose was found to shift in the direction of low temperature as the dose enhanced the demonstration of a second-order kinetics, which could be accredited to the formation of deep-level defects, attributable to the irradiation [590]. The reproducibility tests conducted on the samples demonstrated that the sample preparation in addition to the TL response of the $BaTiO_3$ under gamma irradiation is reproducible and under adequate maximum value. The dose response curve illustrated a good linearity over the dose range observed.

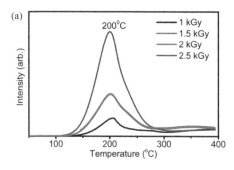 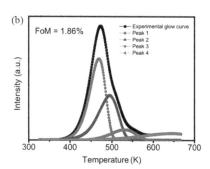

FIGURE 5.2
(a) TL glow curve of Dy^{3+} (4 mol%)-doped $BaTiO_3$ 1 to 2.5 kGy gamma exposure and (b) CGCD curve of $BaTiO_3$:Dy^{3+} (4 mol%)-doped phosphor with 2.5 kGy γ-dose. (Reprinted with the permission from Ref. [593]. Copyright © Springer publications.)

TL glow curve of Dy^{3+}-doped $BaTiO_3$ phosphor gives linear response with γ-dose had outstanding intensity. Kinetic parameters are calculated for different γ-dose-irradiated TL glow curve and estimates that the configuration of deep trapping for γ-irradiated phosphor for the reason that the activation energy is very high ~0.97–1.16 eV for different γ-dose. Moreover, all peaks show the second-order kinetics and high-temperature peaks centered at 200°C with linear response with dose, which demonstrates that the synthesized sample may be beneficial for γ-dosimetry in TL glow curve investigation.

Highly intense TL glow curve is then deconvoluted using the CGCD (computerized glow curve deconvolution) method to calculate trapping parameters [591]. In this proposed work, this computer program (Glow-fit) is based on Halperin and Barner equations which explain the flow of the charges between the various energy levels during a trap unfilled by thermal heating [592]. The trapping parameters of trap levels were indomitable for each deconvoluted peak by this program. The theoretical-produced glow curves were fitted with the experimental glow curves and the quality of fitting was checked by calculating the figure of merit (FOM) for each fitting ascertained by

$$P_s \approx (c/a)^{0.5}$$

Here TL_{exp} and $TL_{Theo.}$ represent the TL intensity of experimental and theoretical glow curves, correspondingly. The summation lengthens in excess of all the accessible experimental data points. Quality of fitting and choice of the proper number of peaks were distinguished by repetition of the procedure of fitting in order to get the minimum FOM with minimum several potential peaks. The fits were considered enough when the FOM values were experiential beneath 5% with most actual being below 2% (Figure 5.2b). In the present examinations, FOM was found to be 1.86%, which substantiates a very good conformity between speculative produced and analytically documented glow curves. Table 5.1 represents the trapping parameters of proposed phosphor calculated by the CGCD method.

TABLE 5.1

Geometrical Factors (μ), Activation Energy (E), and Frequency Factor (s) for γ-Irradiated $BaTiO_3$: Dy^{3+} (4%) Doped Phosphor [594]

Gamma Exposure (KGy)	T_1 (K)	T_m (K)	T_2 (K)	τ	δ	Ω	$\mu = \delta/\omega$	Activation Energy (eV)	Frequency Factors (S^{-1})
1	451	478	505	27	27	54	0.5	1	5.6×10^{22}
1.5	444	474	509	30	35	65	0.53	1.16	5.8×10^{19}
2.5	444	473	501	29	28	57	0.49	0.97	8.08×10^{20}

5.2 Photoluminescence Applications with Specific Titanate Materials

5.2.1 (Ba,Sr)TiO₃:RE Perovskite Phosphors (RE=Dy, Eu): Lighting, Display, and Related Fields

PL emission was observed for fixed concentration of dopant (Dy^{3+}) in $BaTiO_3$ phosphor. It is tracked down to a few particular peaks at visible region; 488 nm (blue emission), 544 nm (green emission), and 621 nm (red emission) had outstanding intensity in PL emission spectra (Figure 5.3a). The emission spectra show attractive characteristics as Dy^{3+} ion in host lattice gives blue, green, yellow, and red emission at the same time. The emission occurs from transition from the $^4F_{9/2}$ level to the ground state and other excited energy levels of Dy^{3+} [595]. Most exceptional peak is found at 544 nm which gives greenish emission. Figure 5.3b shows the CIE coordinate of the Dy^{3+}-activated $BaTiO_3$ material for constant concentration of dopant (4 mol%). The CIE coordinates for the most intense (4 mol% Dy^{3+}-doped $BaTiO_3$) material are estimates that there (0.26, 0.33) which is too closer to the mercantile white light material Y_2O_2S: Eu^{3+} (0.33, 0.33) [596].

Color purity of proposed phosphor was calculated by means of the following formula [597]:

$$3Ca^{2+} \rightarrow 2Eu^{3+} + VCa^{2+}$$

where (X_s, Y_s) are the coordinates of a sample point, (X_d, Y_d) are the coordinates of the dominant wavelength, and (X_i, Y_i) are the coordinates of the

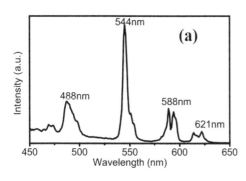

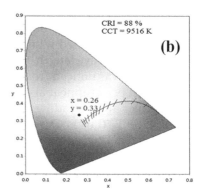

FIGURE 5.3

(a) PL emission spectra of Dy^{3+}-activated $BaTiO_3$ phosphor and (b) Color chromaticity diagram of Dy^{3+}-activated $BaTiO_3$ phosphor. (Reprinted with the permission from Ref. [593] Copyright © Springer publications.)

TABLE 5.2

Transitions for Diverse Peaks in PL Spectra [593]

Peaks	Wavelength	Color	Transitions
1	488 nm	Blue	$^{4}F_{9/2} \rightarrow {}^{6}H_{15/2}$
2	544 nm	Green	$^{4}F_{9/2} \rightarrow {}^{6}H_{13/2}$
3	621 nm	Red	$^{4}F_{9/2} \rightarrow {}^{6}H_{11/2}$

luminiferous point. As per the National Television Standard Committee system, the best white light chromaticity is (0.33, 0.33). It is seen from PL emission spectra that the relative PL intensity especially for green emission intensity is high when contrasted with red and blue emission. CRI value is ~ 88% comprising high CCT of 9516 K for Dy^{3+}-activated $BaTiO_3$ phosphor. Table 5.2 discusses the transitions for diverse peaks in PL spectra.

5.2.2 Yellow–Orange Up-Conversion Emission in Eu^{3+}–Yb^{3+} Co-doped $BaTiO_3$ Phosphor

In $BaTiO_3$: Eu^{3+}–Yb^{3+} sample, the concentration of ytterbium was set at 2.0 mol% while the concentration of europium was shifted from 0.5 to 1.5 mol%. The compositions of the samples taken are displayed beneath

$$2Ca^{2+} \rightarrow Eu^{3+} + K^{+}$$

where $x=0.5$, 0.75, 1, 1.25, and 1.5 mol% and $y=0$, 2.0 mol%

The proposed phosphor was prepared by co-precipitation technique and up-conversion photoluminescence studies were done for this sample. In the proposed work, the up-conversion emission spectra of the $BaTiO_3$: Eu^{3+}–Yb^{3+} phosphors at dissimilar concentrations of Eu^{3+} as displayed in Figure 5.4. The sample was excited with 980 nm diode laser. As Eu^{3+} ion is not able to excite with this wave length, the spectra of co-doped system (Eu^{3+}–Yb^{3+}) are examined. It was noticed from the up-conversion emission spectra that the well-known peaks appear at 489, 592, 614, 654, 704, and 796 nm wavelengths. These peaks are pervading the $^{2}F_{5/2} \rightarrow {}^{2}F_{7/2}$ (Yb^{3+}), $^{5}D_0 \rightarrow {}^{7}F_1$ (Eu^{3+}), $^{5}D_0 \rightarrow {}^{7}F_2$ (Eu^{3+}), $^{5}D_0 \rightarrow {}^{7}F_3$ (Eu^{3+}), $^{5}D_0 \rightarrow {}^{7}F_4$ (Eu^{3+}), and $^{5}D_0 \rightarrow {}^{7}F_6$ (Eu^{3+}) transitions, correspondingly. The highly intense peak caused by Eu^{3+} ion is observed at 614 nm (red region). The inset of Figure 5.4 shows the intensity discrepancy of this peak with Eu^{3+} ion concentration. The highly intense peak is observed for the 0.75 mol% concentration of Eu^{3+}. After this concentration the intensity of all the emission bands is found to diminish with expansion in Eu^{3+} particle concentration. This diminish in intensity occurs because of

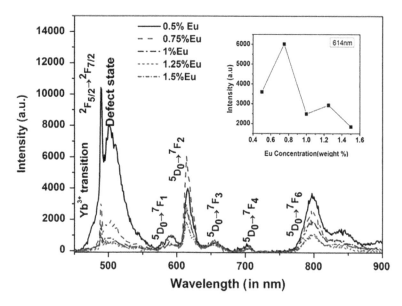

FIGURE 5.4
Up-conversion emission spectra of BaTiO$_3$: Eu^{3+}–Yb^{3+} phosphor at different Eu^{3+} concentrations. (Reprinted with the permission from Ref. [599]. Copyright © Elsevier publications.)

the concentration quenching [165,598]. The effective energy transfer from Yb^{3+} to Eu^{3+} particles is seen at 980 nm excitation of the Yb^{3+} particles. The emission at 614 nm from Eu^{3+} particles is streamlined by changing the Eu^{3+} particles concentration. A piece of Yb^{3+} particles is found to change over into Yb^{2+} state and because of this conversion defect centers are occurred in the host. These defects are found to give expansive emission at around 505 nm. The CIE color coordinates are discovered tunable in a range from green to orange color for this co-doped phosphor demonstrating its appropriateness to be utilized in making the color tunable in display devices (Figure 5.5).

In Eu^{3+}–Yb^{3+} system, the transfer of energy is during cooperative sensitization. Cooperative sensitization is the procedure which includes energy transfer from two excited Yb^{3+} ions penetrating cooperative emission. Initially, the energy of two ytterbium ions unites acquiesce emission at 489 nm which demonstrated in the energy-level diagram by fundamental state. Then the energy of the Yb^{3+} excited states is transferred to the ^{5}D$_1$ excited level of europium ion, which has the analogous energy with the virtual excited level of Yb^{3+}, and then the ^{5}D$_1$ excited states of Eu^{3+} ion relax non-radioactively to the ^{5}D$_0$ level, which is the metastable level from which luminescence is noticed (Figure 5.6). This procedure comprises the concurrent absorption of two photons.

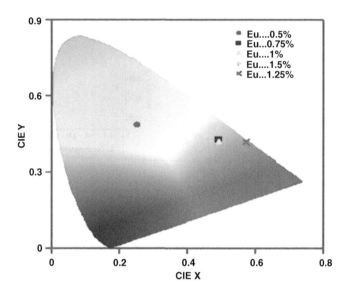

FIGURE 5.5
CIE chromaticity diagram for different Eu^{3+} ion concentration. (Reprinted with permission from Ref. [599]. Copyright © Elsevier publications.)

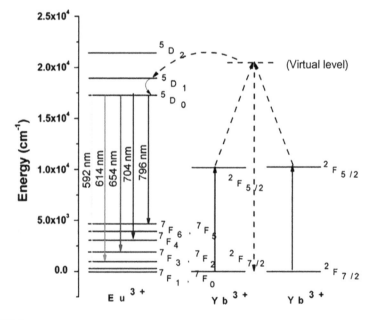

FIGURE 5.6
Energy-level diagram and energy transfer channels in Eu^{3+}–Yb^{3+} phosphor. (Reprinted with the permission from Ref. [599]. Copyright © Elsevier publications.)

5.3 Dielectric and Ferroelectric Properties of Titanates

The current improvement in the efficiency of electronic devices has been leading the research in numerous nanostructured phosphors. The ferroelectric phosphor $BaTiO_3$ has acquired significant awareness in last few years due to its outstanding dielectric and ferroelectric properties. This has escort to noteworthy growth in synthetic techniques that yield high-quality $BaTiO_3$ nanocrystals (NCs) with distinct morphologies (e.g., nanoparticles, nanorods, nanocubes, and nanowires) and controlled crystal phases (e.g., cubic, tetragonal, and multi-phase). The capability to generate nanoscale $BaTiO_3$ with controlled properties facilitates theoretical and experimental investigations on the fascinating yet complex dielectric properties of individual $BaTiO_3$ NCs just as $BaTiO_3$/polymer nanocomposites. Compared with polymer-free individual $BaTiO_3$ NCs, $BaTiO_3$/polymer nanocomposites have a few benefits. The polymeric segment works with basic arrangement processability, high breakdown strength, and light weight for device versatility. The $BaTiO_3$ segment empowers a high dielectric constant. In this part, we feature latest advances in the synthesis of high-quality $BaTiO_3$ NCs by means of a variety of compound advances together with organo-metallic, solvothermal/hydrothermal, template, molten salt, and sol–gel techniques.

Ferroelectricity explains a sample occupying an impulsive electrical polarization that can be exchanged under the utilization of an external electric field. This modifiable polarization is ideal for use in devices for memory storage and microelectronics [600]. Nanoscale ferroelectric samples make it conceivable to plan ferroelectric devices that can function at the atomic scale. Nonetheless, nanoscale ferroelectric materials act in a dissimilar way than their bulk ferroelectric correspondents. The distinctive properties of bulk ferroelectrics narrate to the macroscopic polarization of domains to attain energy depreciation. Absolutely not, nanoscale ferroelectrics acquire constant homogeneous polarization throughout the entirety structure. This fundamentally implies that the part of surface impacts and surface interactions turns out to be more critical. Accordingly, surface impacts play a key role in nanoscale ferroelectricity. The straight piercing of ferroelectricity in un-doped $BaTiO_3$ nanostructures is important for fostering their applications in nanoscale devices and grateful the impacts of size and surface region on the nanoscale [601].

In a two-stage composite material made of high dielectric steady ceramic fillers and a polymer lattice with high breakdown strength, the filler gives the dielectric properties while the lattice gives mechanical properties. Various models have been proposed for anticipating the dielectric properties of polymer–ceramic composites of different syntheses. Contingent upon the clay filler stacking and the polymer and ceramic component interaction, three for the most part acknowledged models have been created for clarifying the noticed properties. These models are simply the Maxwell–Garnett detailing, the Bruggeman self-consistent compelling medium approximation, and the Jaysundere–Smith formulation [602].

The controlled preparation and characterization of $BaTiO_3$ at low measurements, including $BaTiO_3$-containing composites, has impacted the current trend toward the more noteworthy scaling down of $BaTiO_3$-based electronic devices. This review sums up the latest advancement in three key parts of $BaTiO_3$ nanomaterials. First, chemical synthesis techniques for creating excellent $BaTiO_3$ NCs are introduced. Second, the dielectric and ferroelectric properties of $BaTiO_3$ NC clusters and individual $BaTiO_3$ NCs are explored utilizing both hypothetical displaying and piezoresponse force microscopy (PFM). Regardless of the huge advancement made so far in the controlled preparation of BaTiO3 NCs with obvious morphologies (e.g., nanodots, nanorods, nanocubes, and nanowires) and crystal structures (e.g., cubic, tetragonal), they are still of lower quality in contrast with other well-known nanomaterials like metals and semiconductors. For instance, one-dimensional $BaTiO_3$ nanostructures (e.g., nanowires nanorods) still have enormous size disseminations and ineffectively controlled morphologies. Another significant region requiring broad examination is the fitting of the crystal structure construction of $BaTiO_3$ to push its basic size to much more modest qualities. The basic size of $BaTiO_3$ presently detailed is 50 nm, beneath which $BaTiO_3$ NCs are in the cubic phase that needs ferroelectric properties [603–605]. The size impact consequently restricts the use of $BaTiO_3$ in high-density energy/data storage. In view of the core–shell model, the chemistry and structure can likewise impact the ferroelectric behavior of the NCs. Thus, it is feasible to convert the cubic structure into the tetragonal phase at low dimensions through surface designing like ligand alteration and second-stage coating. Different procedures, including ligand adjustment, metal coating, and template confinement, have been demonstrated to be powerful for smothering the presence of the cubic phase [463,606–608]. Be that as it may, these systems are still a long way from seeing use in reasonable applications and a clear understanding of the essential components stays indistinct.

5.4 Lead-Free Relaxor Ceramics Derived from $BaTiO_3$

The progress of leadless ceramics for electrostatic energy storage has fascinated great attention owing to the rising ecological apprehensions. Although there is widespread investigation, the un-success in synergistically optimizing both energy density and efficiency of polycrystalline materials is the major obstacle for their practical applications. In this work, $Bi(Mg_{0.5}Zr_{0.5})O_3$-modified $BaTiO_3$ leadless relax or ferroelectric ceramics are confirmed to be feasible nominees for energy storage (Figure 5.7a). The

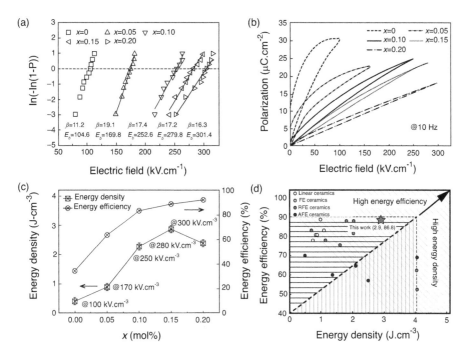

FIGURE 5.7
(a) Weibull plots of dielectric breakdown strengths, (b) polarization hysteresis loops, (c) energy storage properties of (1-x)BT-xBMZ ceramics at ambient temperature, and (d) Energy storage properties of recently reported lead-free ceramics at ambient temperature. (Reprinted with the permission from Ref. [609]. Copyright © Elsevier publications.)

samples can concurrently distribute a high retrievable energy density of 2.9 Jcm^{-3} and a high energy efficiency of 86.8%, which are improved by 625% and 156% over those of original BaTiO$_3$, while holding insensible to thermal incentive over 30–150°C. Figure 5.7b shows the polarization hysteresis loops for (1-x)BT-xBMZ ceramics under the electric field resembling their E_b values at ambience temperature to verify the dissimilarity of ferroelectricity and energy storage properties. Obviously, the pure BT ceramic demonstrates highly saturated loop with a high P_{max} (30.6 µC/cm^2) and a large Pr (11.3 µC/cm^2) at 100 cm/kV, which is a feature of normal ferroelectric but not in favor of providing high energy density and efficiency. On incorporating BMZ into the BT matrix, the polarization hysteresis loop initially transforms into a contracted nonlinear loop (x=0.05), and then turns into sublinear loops (x≥0.10) with a high P_{max} of 17.3–23.2 µC/cm^2 but nearly insignificant Pr of 0.71–1.64 cm^{-2}/µC, which could ensure the fantastic energy density and productivity all the while. Most excitingly, a ultrahigh recoverable energy density of 2.9 J/cm^3, an improvement of 625% over the pure BT ceramics (0.4 J/cm at 100 kV), has been accomplished at x=0.15 at surrounding temperature, joined by a high energy productivity of 86.8%

(see Figure 5.7c). The energy storage properties of recently reported leadless ceramics are summed up in Figure 5.7d, from which it tends to be seen that it is incredibly hard to have high energy density and high energy productivity concurrently.

An ultrahigh energy density of 2.9 J/cm^3 with a high efficiency of 86.8% has been recorded in the environmentally friendly (1-x)BT-xBMZ relaxor ferroelectric ceramics at ambient temperature. It is determined that the addition of BMZ results decreased grain size and improved relative density, which is helpful to develop dielectric breakdown strength. And in situ PFM experiments reveal that PNRs increase the threshold field to induce long-range order and decrease the stability thereof on the nanoscale, in contrast to the typical ferroelectric domains, accounting for the small hysteresis of macroscopic polarization hysteresis for relaxor ferroelectrics. Therefore, the synergistic effects of improved dielectric breakdown strength and dynamic nanoscale domains contribute to the enhanced energy storage properties in (1-x)BT-xBMZ ceramics. Most significantly, the outstanding thermal stability with energy density of 1.77–2.05 J/cm^3 and efficiency of 82.2–89.7% has also been attaining over the wide temperature range of 30°C–150°C. These features endow (1-x)BT-xBMZ lead-free relaxor ferroelectric ceramics with great potential for electrostatic energy storage applications.

5.5 Ceramics with High Permittivity

The constantly rising demands of microelectronic applications have led researchers to be fascinated in searching for new materials with colossal permittivity. In the last few years, $BaTiO_3$ [610], $CaCu_3Ti_4O_{12}$ [611], NiO [612], $Ba(Fe_{0.5}Nb_{0.5})O_3$ [613], (Ba, Sr)TiO_3 [611,614] etc. samples are reported. Regrettably, none of these materials meet the high-energy density storage application necessities. In recent times, a novel colossal permittivity material, (In, Nb) co-doped rutile TiO_2 material, was broadly examined, which possesses high relative permittivity (> 10^4) and low dielectric loss (< 0.05) over a very extensive temperature range from 80 to 450 K [615]. The dielectric properties of other co-doped TiO_2 systems, for example $Ga^{3+}+Nb^{5+}$, $Bi^{3+}+Nb^{5+}$, $Sm^{3+}+Ta^{5+}$, $Al^{3+}+Nb^{5+}$, $Er^{3+}+Nb^{5+}$, $Mg^{2+}+Nb^{5+}$, $Ca^{2+}+Nb^{5+}$, $Zn^{2+}+Nb^{5+}$, $Y^{3+}+Nb^{5+}$, $Mg^{2+}+Ta^{5+}$, and $Sm^{3+}+Nb^{5+}$ co-doped TiO_2 ceramics, are investigated [615–623], and previous experimental results are displayed in Figure 5.8.

In the proposed discussion, two ceramic materials have been discussed namely, $(Eu_{0.5}Nb_{0.5})_xTi_{1-x}O_2$ ceramics and (Sb+Ga) co-doped TiO_2 ceramics, which were prepared by the conventional solid-state method. In $(Eu_{0.5}Nb_{0.5})_x Ti_{1-x}O_2$ ceramics by optimizing the dopant concentration and sintering

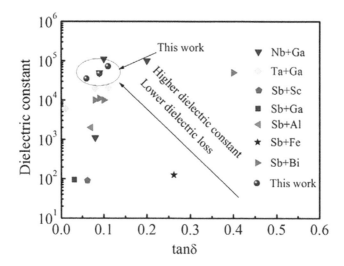

FIGURE 5.8
The dielectric constant and loss of several co-doped TiO_2 ceramics at 1 kHz. (Reprinted with the permission from Ref. [624]. Copyright © Elsevier publications.)

conditions, a low $tan\delta$ (~0.095) and high εr (~2.01×105) are achieved in the $(Eu_{0.5}Nb_{0.5})_{0.1}Ti_{0.99}O_2$ ceramic at room temperature and 1 kHz, which is related to its denser microstructure and smaller grain size distribution. Additionally, the high-temperature relaxation in $(Eu_{0.5}Nb_{0.5})_xTi_{1-x}O_2$ ceramics originates from the electron-hopping motions between Ti^{4+} and Ti^{3+}, while low-temperature relaxation in $(Eu_{0.5}Nb_{0.5})_xTi_{1-x}O_2$ ceramics is owing to dipolar polarization. In addition, the XPS results also verified the existence of defect dipoles and Ti^{3+} ions. Thus, the electron-hopping and electron-pinned defect dipoles should be primarily responsible for the observed high performance colossal permittivity in this work [625]. In (Sb+Ga) co-doped TiO_2 ceramics, the crystalline structure, microstructure, and electrical and dielectric properties of $(Sb_{0.5}Ga_{0.5})_xTi_{1-x}O_2$ ceramics were systematically investigated. XPS revealed the formation of oxygen vacancies, Ti^{3+} and defect dipoles. The enhanced comprehensive dielectric performance with colossal dielectric constant (3.5×104) and low dielectric loss ($tan\ \delta = 0.06$ at 1 kHz and room temperature) have been achieved in the optimum composition with $x = 0.02$. In addition, the dielectric constant shows superior temperature and frequency stability in the range of 25°C–130°C and 20–106 Hz. Impedance results demonstrate that electron pinned defect dipoles (EPDD), internal barrier linear capacitor (IBLC), and electrode effects contribute to the commertially pure (CP) in (Sb+Ga) co-doped TiO_2 ceramics. However, the enhanced interface polarization on the grain boundary and interface between electrode and ceramic will increase the dielectric loss and decrease frequency and temperature stability [624].

5.6 Enhanced Photovoltaic Response

In past few years, perovskite solar cells (PSCs) have been paid more attention owing to its low cost, long charge diffusion length, and high efficiency [626–628]. In 2009, Tsutomu Miyasaka et al. applied perovskite material to solar cells for the first time and obtained the power conversion efficiency (PCE) of 3.8% [629]. Though the PSCs only absorb a small portion of incident light in the visible range, a large portion of incident light energy is lost due to non-absorption of NIR light. This obstruct the further improvement of power conversion efficiency for PSCs [630,631]. One of the strategies to take care of this issue is to extend the absorption light range utilizing up-conversion materials, which can transfer NIR-to-visible light. There have been a few reports on the application of up-conversion materials to dye-sensitized solar cells. There are a few reports on the application of up-conversion materials to PSCs, in which the up-conversion materials applied in these reports are all based on $NaYF_4$. Thus far, there are a small number of reports on the application of TiO_2-based up-conversion materials to PSCs.

5.7 XRD of Perovskite Titanates

X-ray diffraction (XRD) characteristics were observed to examine the crystal structure difference of the five as-quenched films, as demonstrated in Figure 5.9. XRD pattern of combustion-synthesized TiO_2 was recorded in 2θ range from 5 to 50° All the XRD patterns contain numerous intensive peaks, belonging to the fluorine doped tin oxide (FTO), composite perovskite, and an intermediate product. The XRD pattern of the TiO_2 NA/FTO is obtainable in Figure 5.9. The weak peaks, locating at ~26.8° and ~38.2° in Figure 5.9, are accredited to FTO, and the characteristic peaks of TiO_2 are too weak to be observed. Characteristic diffraction peaks due to the trigonal perovskite phase (P3m1) can be observed from all patterns. Furthermore, the peak at ~9.2° belongs to the intermediate product, $MA(FA)I(Br)-PbI_2$-DMSO complex. It tends to be inferred that perovskite crystallites exist together with some intermediate-phase antisolvent quenching. In this, the representative (101) feature was utilized to analyze the perovskite crystal size. Obviously, the crystal structure size increments with expanding the precursor concentration, which likewise includes a quick crystallization in the low precursor concentration samples [632].

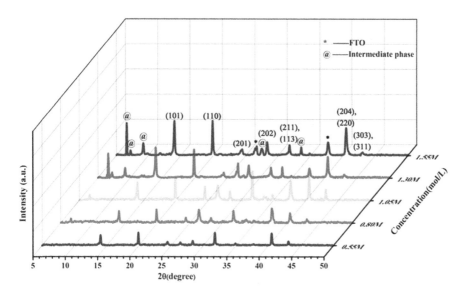

FIGURE 5.9
The XRD patterns of the as-quenched perovskite films from different perovskite precursor concentrations. (Reprinted with the permission from Ref. [632] Copyright © Elsevier publications.)

5.8 Morphology

Perovskite films composed of an infiltrated layer within the TiO_2 nanorod arrays (NAs) and an upper layer are observed to crystallize on the TiO_2 NAs as displayed in Figure 5.10. The thickness of the upper layer increases with increasing the precursor concentration, ranging from 20-nm thick film from a 0.55 M solution to 390 nm thick film from a 1.55 M solution. Assured differences in the perovskite grain can also be pragmatic between the samples from Figure 5.10. Higher precursor concentration leads to larger grains. The mean grain size of the perovskite film increases progressively from 88 to 260 nm when the concentration of the solution is increased from 0.55 to 1.55 M. Undoubtedly such dissimilarity comes from the precursor concentration difference. It is quite significant to realize how the precursor concentration influences the crystallization of the perovskite, which is enlightening for the preparation of high-quality perovskite film. Considering the film-preparation procedure, two stages can be divided as a quenching stage and the annealing stage. Perovskite crystallites along with intermediate-phase complex, $MA(FA)I(Br)-PbI_2-DMSO$, are formed at the first stage. At the second annealing stage, all the intermediate-phase complexes turn into perovskite phase, and all the perovskite crystallites grow up to some amount. To determine which stage is the key process responsible for the crystallization difference,

FIGURE 5.10
The representative (a) morphology and (b) cross-section morphology of PSCs with different perovskite precursor concentrations. As shown in (b), the configuration of all the PSCs is FTO/ TiO_2 compact layer/TiO_2 NA/Perovskite/Spiro-OMeTAD/Ag from down to top. The scale bar is 200 nm. (Reprinted with the permission from Ref. [632] Copyright © Elsevier publications.)

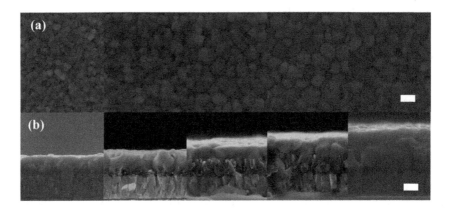

FIGURE 5.11
The surface morphology and cross-section morphology of the as-quenched perovskite films from different precursor concentrations. The scale bar is 200 nm. (Reprinted with the permission from Ref. [632] Copyright © Elsevier publications.)

additional characterization is required. Hence, the SEM images of the surface morphology and the cross-section morphology of the five as-quenched films are presented in Figure 5.11. The mean grain size of the five as-quenched films increases gradually from 75 to 181 nm when the concentration of the solution is increased from 0.55 to 1.55 M. Meanwhile, a slight increase of the perovskite grain size can also be observed at the second annealing stage for all samples. In general, it can preliminarily conclude that the first quenching stage has much more effect on the crystallite difference. These results agree with XRD results.

5.9 Perovskites for Forensic Applications

The RE-doped titanium centered perovskites are considered to be a class of luminescent materials due to intense, sharp, and narrow emission, and its applications are found in various fields, such as sensors, medical, industries, solid-state lightning, and forensics. The use of such luminescent perovskites for visualization of latent fingerprints (LFPs) in forensic science creates much interest, owing to their significant physical and chemical properties. The perovskites for LFPs visualization reveal detailed ridge characteristics on many surfaces, resulting in high contrast, low background hindrance, and high sensitivity and selectivity owing to the strong luminescent emission. Motivated by the above-mentioned idea, several titanium-based luminescent perovskites are employed to visualize LFPs on variety of surfaces, including porous, semi-porous, and non-porous surfaces. As a generic example, Dhanalakshmi et al. synthesized $BaTiO_3$:Eu^{3+} (5 mol%) nanophosphors by using NaCl, NH_4F, and NaBr-assisted solution combustion route [385]. Barium nitrate [$Ba(NO_3)_3$ $4H_2O$ (Sigma 99.9%)], europium nitrate [$Eu(NO_3)_3 5H_2O$ (Sigma 99.9%)], tetra butyl titanate [$Ti(C_4H_9O)_4$ (Sigma 99.9%)], and *Aloe Vera* (*A. V.*) gel were used as initial materials. The noticeable PL intensity enhancement (two-fold) was observed in NH_4F-blended sample. The photometric properties reveal that the $BaTiO_3$:Eu^{3+} (5 mol%), NH_4F (3 wt%) phosphor can emit red color light with high color purity. The optimized nanophosphor was utilized for visualization of LFPs on many surfaces employed via powder-dusting technique. The typical traditional powder-dusting method for LFP development involves collecting the LFPs from various by washing the donor hands neatly with water. Afterward, the washed fingers were nominally pressed on a variety of forensic-related surfaces, including porous and non-porous surfaces. The generated FPs were invisible to naked eye and hence they called latent finger prints. Therefore, to make them visible, the optimized $BaTiO_3$:Eu^{3+} (5 mol%), NH_4F (3 wt%) NPs were dusted gently on the surface with a brushing mode by using a soft feather brush. These developed LFPs were photographed via digital camera under UV 254 nm light. To determine the selectivity of the prepared samples for LFPs visualization, fluxes-assisted prepared NPs were dusted on LFPs impressed on aluminum foil surface and photographed under UV 254 nm light (Figure 5.12a). The obtained results clearly reveal well-defined ridge characteristics developed by using NH_4F flux-blended samples as compared with the other images. In addition, experiment was continued by using many non-porous surfaces, including FPs metal scale, granite stone, stapler, highlighter pen, and metal punch pad. Clear and distinctive ridge patterns were identified, including level 1 (namely delta, loop, and whorl) and level 2 (bifurcation, cross over, lake, hook, short ridge, island, etc.,) with high clarity under 254 nm UV light, owing to its nano-regime and better adhesive nature. However, sharp ridge characteristics were observed on many porous surfaces without

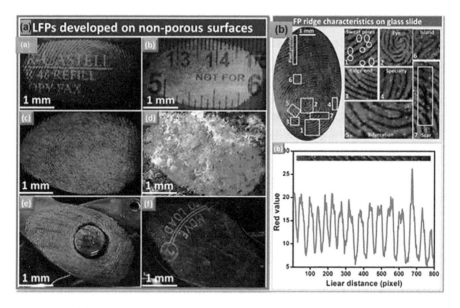

FIGURE 5.12
(a) LFPs visualized on various non-porous surfaces and (b) Various FP ridge characteristics on glass slide stained by using optimized BENF NPs under UV 254 nm light and pixel profile of papillary ridges indicated by rectangle box. (Reprinted with the permission from Ref. [385]. Copyright © Elsevier publications.)

any background interference (Figure 5.12b). The development of LPFs by enabling all level 1–3 ridge details is more essential to identify individuals in advanced forensic science. Dhanalakshmi and co-workers developed the LFPs with three-level ridge details using optimized samples. This result was considered as more vital quantitative data to individualization. In addition, the pixel profile of FP part reveal distinguished ridges (cyan color) and furrows (black color) pattern (Figure 5.12b). Hence, they clearly demonstrated the significance of the $BaTiO_3{:}Eu^{3+}$ (5 mol%), NH_4F (3 wt.%) phosphor in the advanced forensic sciences.

Park and co-workers prepared red-emitting $La_2Ti_2O_7{:}Eu^{3+}$ phosphors via solvothermal reaction method for LFP application [633]. The LFPs of a 33-year old male donor were stained by $La_2Ti_2O_7{:}Eu^{3+}$ phosphors using a soft feather brush. The developed FPs were photographed using DSLR Canon EOS 100D with 5 mm lens (SIGMA MACRO, 50 mm, F2.8, EXDG) under 254 nm wavelength. As compared with sizes of the narrower ridge (normally 427/483 μm) and sweat pore (~ 88–220 μm) [634], the prepared $La_2Ti_2O_7{:}Eu^{3+}$ phosphors are extremely small, as evident from SEM images. This demonstrated the applicability of the prepared nanophosphor for latent fingerprint visualization. The practical applicability of the prepared sample for visualization of LFPs on various surfaces was clearly demonstrated. The clear images of well-defined ridges on various

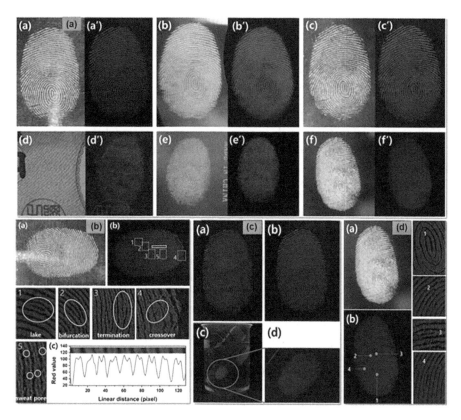

FIGURE 5.13
(a) LFPs developed on various substrates with $La_2Ti_2O_7$:$3Eu^{3+}$phosphors under normal and the 254 nm illumination in dark field, (b) Images of LFPs developed on aluminum foil detected under bright field and dark field, showing various ridge details and variation of the red value over a few papillary ridges, (c) LFPs images visualized by pristine sample, aged sample for 1 week, and in the water, and (d) LFPs developed on transparent plastic sheet under normal and UV 254 nm light. (Reprinted with permission from Ref. [633] Copyright © Elsevier publications.)

substrates without background interference were exhibited (Figure 5.13a). Generally, LFPs contain 99% water, and many organic/inorganic salts (such as chloride and phosphate) remain on the LFPs up to several days [635]. These available salts show excellent adherence properties which bind the $La_2Ti_2O_7$:Eu^{3+} powder, indicating that the prepared powder showed excellent hydrophilic nature, which is a necessary requirement for forensic science sample collection. The detailed defined ridge flow, ridge oriental field, and ridge pattern types were clearly demonstrated on aluminum foil under UV 254 nm light illumination. However, the level 1 details which are not sufficient for personalization was revealed without UV illumination. The magnified images showed detailed level 2 features (bifurcation, lake, crossover, and termination), and level 3 features (sweat pore) are clearly observed, which are useful in the identification of partial or damaged fingerprints (Figure 5.13b). The red value in the pixel profile exhibited difference between bright field of ridges and dark field of furrow patterns under UV illumination (Figure 5.13b). Furthermore, the developed FPs after immersion in water reveal small background interferences, but it shows a well-resolved image, which exhibits the physical and chemical stability of the prepared nanophosphors (Figure 5.13c). In magnified images (Figure 5.13d), the level 2 features such as origination, termination, bifurcation, and lake were clearly distinguished under UV illumination suggesting that the prepared $La_2Ti_2O_7$:3 mol% Eu^{3+} phosphor was considered to be a potential candidate for the LFP detection.

Similarly, Saif et al. [636] prepared $La_2Ti_2O_7$:Eu^{3+} and dispersed in silica matrix phosphor powder by sol–gel process. The powder-dusting technique was employed to visualize LFPs on many surfaces. The prepared $La_2Ti_2O_7$:Eu^{3+} (10 mol%) which dispersed in silica matrix in the presence of 0.1 mol cytiltrimethyl ammonium bromide (CTAB) shows long lifetime and high-intense red luminescence. Hence, this optimized sample was utilized as a labeling material for LFP visualization. The obtained results clearly showed the fresh fingerprint images on glass, aluminum foil, opaque plastic pages, transparent-colored plastic bags, and highly scratched CDs owing to characteristic small size of the sample (Figure 5.14a). The exhibited results endorse that prepared nanophosphor is a useful fingerprint-labeling owing to its strong PL and photo-stable property. In addition, the same group of workers, succeeded in production of high PL, photo-stable and non-toxic $Y_2Ti_2O_7$:Eu^{3+} (1, 2, 4, 6, 10, 15 and 30 mol%) nanophosphor embedded into silica matrix via sol–gel route [637]. The in vivo toxicity test of the prepared nanophosphor embedded into silica matrix exhibits non-toxic (LD_{50}) for $Y_2Ti_2O_7$:Eu^{3+} (15 mol%). Hence, the prepared nanophosphor was successfully used to develop LFPs on various forensic relevant materials, including non-porous and porous surfaces. The fresh clear finger print images were visualized by $Y_2Ti_2O_7$:Eu^{3+} nanophosphor embedded into silica matrix on various surfaces, including glass, aluminum foil, colored plastic bags, high scratched CD, blue paper, and freshly cut green leaves due to the characteristic small size of the nanophosphors as well as excellent adhesion property of the material via both electrostatic and adsorption interactions and increase the chemical stability, allowing the long-term preservation of fluorescence and affording affinity with LFPs. The semi-porous colored paper and porous surface of freshly cut green leaf represent one of the difficult, but frequently encountered, substrates for recovering LFPs. A clear contrast between the fingerprint ridges and the background was also visible. The aged uncovered fingerprints on the glass, aluminum foil, yellow plastic page, and CD were developed after 7 days with clear details (Figure 5.14b). In 2018, Park et al. [638] reported that the orange-red-emitting Eu^{3+} ions activated cubic $Gd_2Ti_2O_7$ nanophosphors through a citrate sol–gel method. The increment in PL intensity with increasing Eu^{3+} ion concentration indicated an optimum concentration of 0.09 mol.

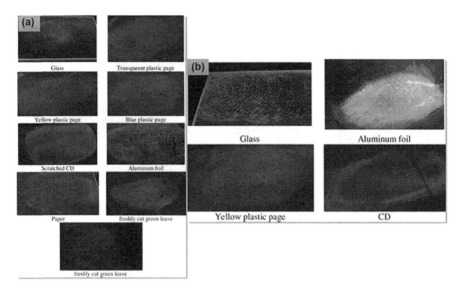

FIGURE 5.14
Photographs of fingerprints developed using $Y_2Ti_2O_7$:Eu^{3+}nanophosphor embedded into silica matrix on (a) many surfaces and (b) aged fingermarks for 7 days from different surfaces excited with a mercury lamp. (Reprinted with permission from Ref. [637] Copyright © Elsevier publications.)

The emission spectra appeared on orange-red emission ($^5D_0 \rightarrow {}^7F_1$) with the CIE coordinates of ($x=0.615$, $y=0.384$).

The experiments to visualize LFPs collected from different persons stained by $Gd_2Ti_2O_7$:$0.09Eu^{3+}$ on the glass surface are performed without and with UV light (Figure 5.15a). The results clearly showed the improved contrast and resolution of FPs images, which signifies the prepared material as a labeling agent for easy identification of individuals. Furthermore, the practical applicability of the prepared phosphor for LFP visualization on many surfaces was clearly demonstrated (Figure 5.15b). The obtained level 2 (ridge termination, hook, bifurcation, core, and delta) as well as level 3 details on aluminum foil under 365 nm demonstrate the potential application of $Gd_2Ti_2O_7$:$0.09Eu^{3+}$ for LFP visualization (Figure 5.15c).

Sandhyarani et al. [381] prepared SiO_2@$SrTiO_3$:Eu^{3+} (1 mol%), Li (1 wt%) core–shell nanopowders via combustion process to overcome the shortcomings encountered in enhancing the quality of LFPs by traditional powder-dusting method. The detailed schematic representation of visualization of LFPs on various surfaces using optimized SiO_2@$SrTiO_3$:Eu^{3+} (1 mol%), Li (1 wt%) NPs was proposed (Figure 5.16a). The results clearly showed that the SiO_2-coated $SrTiO_3$:Eu^{3+} particles with various cycles reveal clear FP images with necessary minutiae details, which are highly necessary during crime spot investigations. All three levels of LFP ridge characteristics comprising

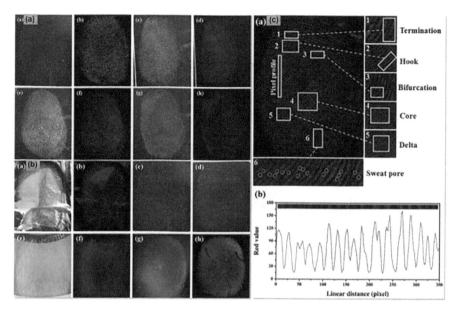

FIGURE 5.15
Digital photographs of LFPs visualized using $Gd_2Ti_2O_7$:0.09Eu^{3+} (a) collected from different persons on the glass surface, (b) on different substrate surfaces with and without UV light, and (c) optical microscope images for level 2–3 FP detection on aluminum foil surface and fluctuations of red value with ridge (red) and furrow (black) over a few papillary ridges. (Reprinted with the permission from Ref. [638] Copyright © Elsevier publications.)

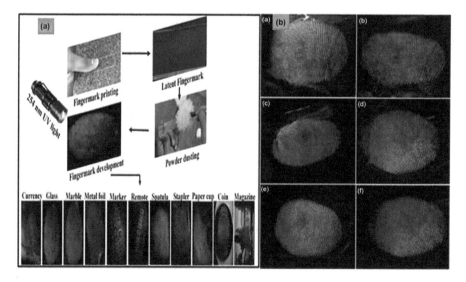

FIGURE 5.16
(a) Schematic representation of visualization of LFPs on various surfaces and (b) LFPs visualized by staining $SrTiO_3$:Eu^{3+} (1 mol%), Li (1 wt%) NPs with different SiO_2 coat cycles. (Reprinted with permission from Ref. [381] Copyright © Elsevier publications.)

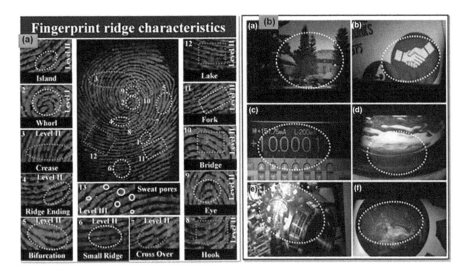

FIGURE 5.17
(a) Various ridge details of LFPs visualized using optimized NPs on glass surface under UV 254 nm and (b) Visualized LFPs by staining SiO_2@$SrTiO_3$:Eu^{3+} (1 mol%), Li (1 wt%) NPs by powder-dusting method on porous surfaces. (Reprinted with the permission from Ref. [381] Copyright © Elsevier publications.)

core (level 1), ridge ending (level 2), bifurcation (level 2), island/dot (level 2), scar (level 3), and pore (level 3) were established by means of dusting method, which validated its possibility for LFP detection. The developed LFPs on various porous materials, such as various types of paper, including train tickets, printing papers, magazine covers, and note papers exhibit clear ridge details without any background interference, due to high florescence property of the prepared sample (Figure 5.16b). The three-level ridge characteristics comprising core, ridge ending, bifurcation, island/dot, scar, and pore were clearly revealed (Figure 5.17a). In addition, authors a performed series of experiments after the FPs had been aged for several time periods. It was noticed that the detection sensitivity diminishes slowly with extended aging of the FPs, owing to the slow evaporation of the LFP. However, the aged LFPs (four months) could be revealed with necessary ridge details, indicating that the prepared samples were high enough to visualize aged LFPs with high sensitivity. In addition, the developed LFPs using prepared sample on porous surfaces showed all three level details as compared with commercially used magnetic materials (Figure 5.17b), demonstrating the efficiency of prepared red-emitting phosphor.

In 2018, Dhanalakshmi et al. [370] synthesized the superstructures of europium ions-doped $BaTiO_3$ nanophosphors (NPs) via sonochemical route using *A. V.* gel as a bio-template for the visualization of LFPs. The technical details of LFP visualization using $BaTiO_3$: Eu^{3+} (5 mol%) NPs by powder-dusting process were discussed. As compared with conventionally used micron-sized magnetic, bronze, and yellow fluorescent powder, the opti-mized $BaTiO_3$: Eu^{3+} (5 mol%) NPs exhibits extended minutiae level 2 ridge patterns effortlessly and effectively due to their smaller crystallite size. From the obtained results, it was confirmed that, prepared NPs were considered to be a new probe for visualization of LFPs on various surfaces due their strong luminescence property. In addition, the aged LFPs on glass showed progressively decreased sensitivity as prolonged aging of the LFPs, due to evaporation of FPs compositions. Nevertheless, level 2 sharp ridge minutiae of the LFPs aged almost 1 month can be effortlessly detected, demonstrating the practicability of the optimized sample. The versatility of the prepared sample for LFP visualization on different porous surfaces namely magazine covers and Indian currency with dissimilar background colors were evalu-ated (Figure 5.18a). Interestingly, high contrast, sufficient quality clear ridge patterns on all surfaces were evidently determined without any background interference. The post-processed FP images on aluminum foil showed level 2 and level 3 (sweat pores) minutiae ridges effortlessly due to their smaller crys-talline size (Figure 5.18b). Furthermore, the same group of workers prepared broom-like hierarchical structures of $BaTiO_3$:Nd^{3+} (1–11 mol%) NPs via CTAB-assisted solvothermal route [639]. LFPs were visualized on non-porous sur-faces, such as highlighter, marker, spatula, CD, glass slide, TV remote, battery charger, and spray bottle by staining optimized $BaTiO_3$:Nd^{3+}(7 mol%) NPs under 254 nm UV light. Well-defined ridge details including all three levels

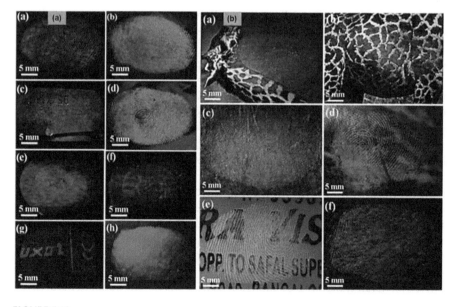

FIGURE 5.18
LFPs under UV 254 nm visualized by $BaTiO_3$:Eu^{3+} (5 mol%) NPs on various (a) non-porous sub-strate surfaces and (b) paper substrates with different background colors. (Reprinted with per-mission from Ref. [370] Copyright © Elsevier publications.)

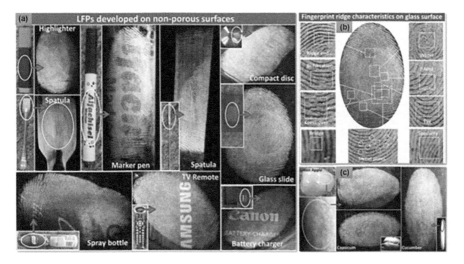

FIGURE 5.19
LFPs under UV 254 nm visualized by $BaTiO_3:Nd^{3+}$ (7 mol%) NPs on various (a) non-porous substrate surfaces, (b) various ridge minutiae, and (c) fruits and vegetables. (Reprinted with permission from Ref. [639] Copyright © Elsevier publications.)

were clearly visualized due to their nano-regime and better adhesion nature (Figure 5.19a). LFP development experiments were performed by considering many porous surfaces, including playing cards, magazine, spray bottle surfaces to determine background hindrance of optimized NPs under UV 254 nm light. The obtained result shows clear FPs without any background interference after UV light irradiation. In addition, all the three levels (whorl, ridge end, bi-furcation, crossover, eye, island, and sweat pores) of ridge details were clearly visualized on glass surface (Figure 5.19b). Furthermore, sensitivity and selectivity of prepared NPs for LFP visualization were evaluated by considering most neglected fruits and vegetable surfaces, such as green apple, capsicum, and cucumber under 254 nm light (Figure 5.19c). Normally, fruits and vegetables subjected to deterioration depends on environmental and storage conditions (e.g., temperature and humidity). The defined ridges were revealed on the apple and capsicum surfaces, owing to its smooth and longer shelf life. However, partial LFPs were visualized on the surface of cucumber. Similarly, glassy and smooth-surfaced tomato shows less clarity LFPs due its color interference. On the other hand, visualization of LFPs on the onion and potato surfaces was achieved using optimized NPs, but only level 1 detail can be enhanced due their contrast colors. From overall obtained results, it was evident that the prepared sample and followed technique was versatile for the development of LFPs with excellent sensitivity, selectivity, and without background hindrance.

In 2017, Dhanalakshmi et al. [367] synthesized $BaTiO_3:Dy^{3+}$ (2 mol%) hollow/solid microspheres by ultrasonication method. The aged (different time periods) LFPs were visualized by using optimized NPs via powder-dusting

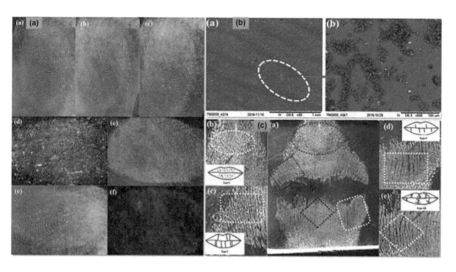

FIGURE 5.20
LFPs under UV 254 nm visualized by $BaTiO_3:Dy^{3+}$ (2 mol%) NPs on various (a) textured marble stones, (b) magnified SEM images of fingerprints, and (c) Lip print and its different grooves. (Reprinted with permission from Ref. [367] Copyright © Elsevier publications.)

technique to determine the suitability and robustness of the prepared powder. The obtained photographs showed defined ridge substructures in even one-month aged LFPs, which demonstrate that the sensitivity of labeling powder linearly decreases as aging of the LFPs enhances, due to evaporation of chemical constituents of the LFPs. In addition, visualized LFPs on various textured marble stones by optimized $BaTiO_3:Dy^{3+}$ (2 mol%) powder under UV 254 nm light validated the well-defined ridge characteristics with excellent contrast and without or less background hindrance (Figure 5.20a). Furthermore, magnified SEM images of developed LFP shows uniformly distributed particles with stronger adhesive ability via each static and surface absorption interactions (Figure 5.20b). However, lip prints similar to fingerprints have many elevations and depressions, which provide evidential information for individualization. Normally, lip prints can be found where the surface is in contact with the lips, but are invisible to naked eye. Hence, much effort has been necessary to make them visible. Hence, authors demonstrated the visualization of lip prints on glass surface using $BaTiO_3:Dy^{3+}$ (2 mol%) powder followed by powder-dusting method (Figure 5.20c). The obtained image clearly showed whole lip prints with Tsuchihashi's Type V, Type I, Type II, and Type III grooves with high sensitivity and contrast due to nano-regime and adhesive property of the powder.

Similarly, Girish et al. [640] explored the superstructures of dysprosium (Dy^{3+})-doped Zn_2TiO_4 via a facile solution combustion route. The LFPs visualized by staining white luminescent nanophosphor showed an excellent contrast to the finger mark ridges without any background hindrance, due

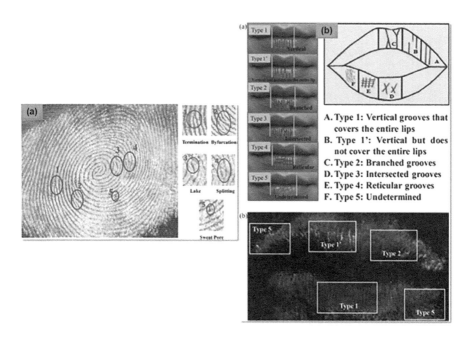

FIGURE 5.21
(a) LFPs visualized using $Zn_2TiO_4:Dy^{3+}$ NPs under UV 254 nm on glass surface and (b) Patterns of lip prints and various lip print patterns identified using $Zn_2TiO_4:Dy^{3+}$ NPs. (Reprinted with permission from Ref. [640] Copyright © Elsevier publications.)

to nano size as confirmed by TEM. The visualized FPs comprised various ridge substructures, including ridge termination, ridge splitting, crossover, and lake bifurcation which are unique to identify individuals (Figure 5.21a). However, lip prints were normally categorized based on geometric dominance of lines in the prints, such as vertical, intersected, branched, reticular, as well as undetermined patterns. Generally, female lip prints comprised of more vertical and intersected patterns while branched and reticular patterns are predominant in males. The authors successfully developed lip print using optimized powder and identified various categories of patterns, and the obtained result showed more vertical and intersected grooves indicating the female dominant nature (Figure 5.21b).

6

Challenges Ahead

Perovskites have demonstrated initial assurance in their applications for light emission. Be that as it may, much stays to be done before these materials can understand a definitive objective of persistent wave and, preferably, electrically energized lasing. Expanding the span of the excitation pulse that produces sustained inversion and lasing, eventually route for continuous-wave operation, is initially pushed. An essential requirement for applications of perovskite photonic sources will be stability under the relevant injection conditions. Light-emitting perovskite devices face many of the stability concerns seen in perovskite solar cells, perhaps especially severely as a result of the high carrier densities typically generated in these devices.

The hazardness of perovskite precursor materials is also a significant point of contemplation. All of the perovskites used so far for light-emitting applications, much like any high-performance perovskite solar cell, contain lead, a noxious material. Applying proper end of life management of any light-harvesting or light-emitting device using lead-based perovskites is critical to mitigating environmental impact. Incineration of perovskite films and subsequent recovery of the lead has been determined to be the most environmentally sustainable protocol. Studies on lead-free perovskites for light emission are a hopeful yet unfamiliar opportunity.

7

Concluding Remarks and Future Perspectives

In the beginning of this chapter, we describe state of the art of perovskite materials, its unique structure, and its uses in different types of energy materials and storage devices. Here, we are giving a brief introduction of development of TiO_2-based materials, namely titanate and titania, for environmental and energy applications. The titanate and titania materials can be easily transformed to each other via simple processes. For instance, titanate materials can be transformed to titania via thermal annealing or hydrothermal treatment of hydrogen titanate, and the titania material can react with the alkaline solutions to form titanate materials. A clear advantage of synthesizing titania from titanate is to maintain or make use of the various nanostructures easily accessible for titanate materials. Owing to the unique crystal structure and physical-chemical properties, titanate materials show great potential for applications in adsorption, lithium ion batteries, biomedical applications, self-cleaning, and oil-water separation. Since titanate materials possess large interlayer spacing and adjustable lattice parameters, they can be widely used as adsorbents due to their unique ion-exchange property. To be a good adsorbent, the adsorption capacity and removal rate are important consideration. The adsorption performance depends on the surface area and the number of exchangeable cations in the titanate materials.

For the energy storage application, titanate and titania materials show good potential in replacing the commercial carbon negative electrodes, offering excellent rate capacities and ultralong cycle life with enhanced safety benefited from the high lithiation potential, and absence of the lithium electroplating and the charge loss problem arising from the formation of a solid-electrolyte interface layer. As an important photoanode for the DSSC and perovskite solar cells, the engineering of TiO_2 nanostructures toward prolonging the charge carrier lifetime, and bandgap matching with dye (organo-metal halides), are critical to achieving the goal of high efficiency and low cost in the utilization of solar energy. By making use of the reversible transition from Ti^{4+} to Ti^{3+} into TiO_2 host by ion intercalation/deintercalation, the visible electrochromism effect can be achieved on the TiO_2 materials. For the photocatalytic application, due to their wide band gap and the crystal defects, the photocatalytic performance of titanate materials is not as good as TiO_2. Thus, it is essential to obtain the titania materials with required crystal structure and designed nano-construction from titanate by rational control of the morphology of the starting material. An interesting strategy is to

DOI: 10.1201/9781003381907-7

combine the excellent adsorption ability of titanates and high photocatalytic activity of titania to promote faster photo degradation of organic pollutants in wastewater. Also, the doping technology and hybrid structures are also introduced to enhance the visible light activity.

Future effort should be focused on the improvement of its functionalities through materials and structural optimization, eventually leading to practical environmental, energy and other related applications. Practical application of titanate and titania materials is realistically achievable as these materials are intrinsically stable in a wide range of chemical media and under UV illumination, earth-abundant and low cost in production. For the absorbent application, the mass production on titanate materials with high surface area is an important future effort toward low-cost and effective pollutant removal, especially for the toxic metal ions and nondegradable organic dye. Little attention has been paid so far on the disposal or recovery of the titanate materials after adsorption. More research effort should be made in this aspect of the future. As a well-studied photo catalyst system in the past decades, the photocatalytic efficiency of TiO_2 under full solar energy spectrum is still low. New photo conversion concept based on rational materials design or bandgap engineering is urgently needed to achieve high solar-to-fuel conversion efficiency. For other energy-related applications, much effort is needed to commercialize these TiO_2-based materials for rechargeable lithium ion battery and DSSC applications. In particular, the exploration on the TiO_2 materials toward solid state, wearable and printable energy devices will have a promising future for the development of efficient and flexible optoelectronics. For the wetting ability control on TiO_2 nanostructure, the rate of wetting property change, stability under UV and mechanical strength should be carefully studied in the future.

References

1. Ahmadzai S, McKinna A. Afghanistan electrical energy and trans-boundary water systems analyses: Challenges and opportunities. *Energy Rep* 2018;4:435–69. https://doi.org/10.1016/j.egyr.2018.06.003.
2. Rivotti P, Karatayev M, Mourão ZS, Shah N, Clarke ML, Dennis Konadu D. Impact of future energy policy on water resources in Kazakhstan. *Energy Strategy Rev* 2019;24:261–7. https://doi.org/10.1016/j.esr.2019.04.009.
3. Zhang R, Peng M, Zheng C, Xu K, Hou X. Application of flow injection–green chemical vapor generation–atomic fluorescence spectrometry to ultrasensitive mercury speciation analysis of water and biological samples. *Microchem J* 2016;127:62–7. https://doi.org/10.1016/j.microc.2016.02.006.
4. Veldhuis SA, Boix PP, Yantara N, Li M, Sum TC, Mathews N, et al. Perovskite materials for light-emitting diodes and lasers. *Adv Mater* 2016;28:6804–34. https://doi.org/10.1002/adma.201600669.
5. Kabir E, Kumar P, Kumar S, Adelodun AA, Kim KH. Solar energy: Potential and future prospects. *Renew Sustain Energy Rev* 2018;82:894–900. https://doi.org/10.1016/j.rser.2017.09.094.
6. Knör G. Recent progress in homogeneous multielectron transfer photocatalysis and artificial photosynthetic solar energy conversion. *Coord Chem Rev* 2015;304–305:102–8. https://doi.org/10.1016/j.ccr.2014.09.013.
7. Sundseth K, Pacyna JM, Pacyna EG, Pirrone N, Thorne RJ. Global sources and pathways of mercury in the context of human health. *Int. J. Environ. Res. Public Health* 2017;14(1):105. https://doi.org/10.3390/ijerph14010105.
8. Mukherjee AB, Zevenhoven R, Brodersen J, Hylander LD, Bhattacharya P. Mercury in waste in the European Union: Sources, disposal methods and risks. *Resour Conserv Recycl* 2004;42:155–82. https://doi.org/10.1016/j.resconrec.2004.02.009.
9. Maggio G, Nicita A, Squadrito G. How the hydrogen production from RES could change energy and fuel markets: A review of recent literature. *Int J Hydrogen Energy* 2019;44:11371–84. https://doi.org/10.1016/j.ijhydene.2019.03.121.
10. Tanaka H, Misono M. Advances in designing perovskite catalysts. *Curr Opin Solid State Mater Sci* 2001;5:381–7. https://doi.org/10.1016/S1359-0286(01)00035-3.
11. Arandiyan H, Wang Y, Sun H, Rezaei M, Dai H. Ordered meso- and macro-porous perovskite oxide catalysts for emerging applications. *Chem Commun* 2018;54:6484–502. https://doi.org/10.1039/c8cc01239c.
12. Saparov B, Mitzi DB. Organic-inorganic perovskites: Structural versatility for functional materials design. *Chem Rev* 2016;116:4558–96. https://doi.org/10.1021/acs.chemrev.5b00715.
13. Bin Adnan MA, Arifin K, Minggu LJ, Kassim MB. Titanate-based perovskites for photochemical and photoelectrochemical water splitting applications: A review. *Int J Hydrogen Energy* 2018;43:23209–20. https://doi.org/10.1016/j.ijhydene.2018.10.173.
14. van Sark WG, de Wild J, Rath JK, Meijerink A, Schropp RE. Upconversion in solar cells. *Nanoscale Res Lett* 2013;8:1–10. https://doi.org/10.1186/1556-276x-8-81.
15. García-López EI, Marcì G, Palmisano L. Photocatalytic and catalytic reactions in gas–solid and in liquid–solid systems. *Heterogeneous Photocatalysis* 2019:153–76. https://doi.org/10.1016/b978-0-444-64015-4.00005-5.

16. Senocrate A, Maier J. Solid-state ionics of hybrid halide perovskites. *J Am Chem Soc* 2019;141:8382–96. https://doi.org/10.1021/jacs.8b13594.

17. Garcia R, Cervera R. Morphology and structure of $Ni/Zr_{0.84}Sc_{0.16}O_{1.92}$ electrode material synthesized via glycine-nitrate combustion method for solid oxide electrochemical cell. *Appl Sci* 2019;9:264. https://doi.org/10.3390/app9020264.

18. Kumar A, Kumar A. Electrochemical behavior of oxygen-deficient double perovskite, $Ba_2FeCoO_{6-\delta}$, synthesized by facile wet chemical process. *Ceram Int* 2019;45:14105–10. https://doi.org/10.1016/j.ceramint.2019.04.110.

19. Shan X, Zhang S, Zhou M, Geske T, Davis M, Hao A, et al. Porous halide perovskite–polymer nanocomposites for explosive detection with a high sensitivity. *Adv Mater Interfaces* 2019;6:1–7. https://doi.org/10.1002/admi.201801686.

20. Singh D, Tabari T, Ebadi M, Trochowski M, Yagci MB, Macyk W. Efficient synthesis of $BiFeO_3$ by the microwave-assisted sol-gel method: "A" site influence on the photoelectrochemical activity of perovskites. *Appl Surf Sci* 2019;471:1017–27. https://doi.org/10.1016/j.apsusc.2018.12.082.

21. Zhang W, Lu Y, Du H, Long J. Author's accepted manuscript. *Ceram Int* 2016. https://doi.org/10.1016/j.ceramint.2016.10.043.

22. Liu C-Y, Tsai S-Y, Ni C-T, Fung K-Z, Cho C-Y. Enhancement on densification and crystallization of conducting $La_{0.7}Sr_{0.3}VO_3$ perovskite anode derived from hydrothermal process. *Jpn J Appl Phys* 2019;58:SDDG03. https://doi.org/10.7567/1347-4065/ab0df0.

23. Panse VR, Kokode NS, Dhoble SJ, Yerpude AN. Luminescence investigation of novel MgPbAl10O17:Tb3+ green-emitting phosphor for solid-state lighting. *Luminescence* 2016;2015:893–6. https://doi.org/10.1002/bio.3049.

24. Marchant BP, Saby NPA, Arrouays D. A survey of topsoil arsenic and mercury concentrations across France. *Chemosphere* 2017;181:635–44. https://doi.org/10.1016/j.chemosphere.2017.04.106.

25. Singh SN, Singh AN, Singh TK, Singh TR. Analysis of the glow curves of natural salt by thermoluminescence technique. *Int. J. Lumin. Appl* 2016:28–31.

26. Dubey V, Kaur J, Parganiha Y, Suryanarayana NS, Murthy KVR. Study of formation of deep trapping mechanism by UV, beta and gamma irradiated Eu^{3+} activated SrY_2O_4 and $Y_4Al_2O_9$ phosphors. *Appl Radiat Isot* 2016;110:16–27. https://doi.org/10.1016/j.apradiso.2015.12.047.

27. K. Sheu et al. White-light emission from near UV InGaN-GaN LED chip precoated with blue/green/red phosphors. *IEEE Photonics Technology Letters* Jan. 2003; 15, no. 1:18–20. https://doi.org/10.1109/LPT.2002.805852.

28. Höök M, Tang X. Depletion of fossil fuels and anthropogenic climate change-A review. *Energy Policy* 2013;52:797–809. https://doi.org/10.1016/j.enpol.2012.10.046.

29. Ajibola, O O E, Ibidapo-obe O, Balogun OJ. Developing sustainable renewable energy for rural dwellers' energy sufficiency. *ABUAD J Eng Res Dev* 2017;1:1–7.

30. Bicer Y, Dincer I. Life cycle environmental impact assessments and comparisons of alternative fuels for clean vehicles. *Resour Conserv Recycl* 2018;132:141–57. https://doi.org/10.1016/j.resconrec.2018.01.036.

31. De Marco Nicholas. *Defect and Grain Boundary Engineering for Enhanced Performances and Lifetimes of Hybrid Perovskite Solar Cells*. University of California, Los Angeles ProQuest Dissertations Publishing, 2019, 13887239.

32. Jain S, Jain NK, Vaughn WJ. Challenges in meeting all of India's electricity from solar: An energetic approach. *Renew Sustain Energy Rev* 2018;82:1006–13. https://doi.org/10.1016/j.rser.2017.09.099.

33. Dewangan P, Bisen DP, Brahme N, Tamrakar RK, Upadhyay K, Sharma S, et al. Studies on thermoluminescence properties of alkaline earth silicate phosphors. *J Alloys Compd* 2017;735:1383–88. https://doi.org/10.1016/j.jallcom.2017.11.293.

34. Paerl HW, Havens KE, Hall NS, Otten TG, Zhu M, Xu H, et al. Mitigating a global expansion of toxic cyanobacterial blooms: Confounding effects and challenges posed by climate change. *Mar Freshw Res* 2019. https://doi.org/10.1071/MF18392.

35. Aryanpur V, Atabaki MS, Marzband M, Siano P, Ghayoumi K. An overview of energy planning in Iran and transition pathways towards sustainable electricity supply sector. *Renew Sustain Energy Rev* 2019;112:58–74. https://doi.org/10.1016/j.rser.2019.05.047.

36. Marselle MR, Stadler J, Korn H, Irvine KN, Bonn A. *Biodiversity and Health in the Face of Climate Change: Perspectives for Science, Policy and Practice.* 2019. ISBN 978-3-030-02318-8. https://doi.org/10.1007/978-3-030-02318-8_20.

37. Wen D, Kuwahara H, Kato H, Kobayashi M, Sato Y, Masaki T, et al. Anomalous orange light-emitting (Sr, Ba)$_2$SiO$_4$: Eu^{2+} phosphors for warm white LEDs. *ACS Appl Mater Interfaces* 2016;8:11615–20. https://doi.org/10.1021/acsami.6b02237.

38. Streets DG, Horowitz HM, Lu Z, Levin L, Thackray CP, Sunderland EM. Global and regional trends in mercury emissions and concentrations, 2010–2015. *Atmos Environ* 2019;201:417–27. https://doi.org/10.1016/j.atmosenv.2018.12.031.

39. Benasla M, Hess D, Allaoui T, Brahami M, Denaï M. The transition towards a sustainable energy system in Europe: What role can North Africa's solar resources play? *Energy Strategy Rev* 2019;24:1–13. https://doi.org/10.1016/j.esr.2019.01.007.

40. Wu X, Zheng J, Ren Q, Zhu J, Ren Y, Hai O. Luminescent properties and energy transfer in novel single-phase multicolor tunable Sr$_3$Y(BO$_3$)$_3$: Tb^{3+}, Eu^{3+} phosphors. *J Alloys Compd* 2019;805:12–8. https://doi.org/10.1016/j.jallcom.2019.07.061.

41. Sánchez A, Martín M. Optimal renewable production of ammonia from water and air. *J Clean Prod* 2018;178:325–42. https://doi.org/10.1016/j.jclepro.2017.12.279.

42. Mukai H, Tamura K, Kikuchi R, Takahashi Y, Yaita T, Kogure T. Cesium desorption behavior of weathered biotite in Fukushima considering the actual radioactive contamination level of soils. *J Environ Radioact* 2018;190–191:81–8. https://doi.org/10.1016/j.jenvrad.2018.05.006.

43. Yang Y, Li J, Liu B, Zhang Y, Lv X, Wei L, et al. Synthesis and luminescent properties of Eu^{3+}, Eu^{3+}/Bi^{3+} and Gd^{3+} codoped YAG: Ce^{3+} phosphors and their potential applications in warm white light-emitting diodes. *Chem Phys Lett* 2017;685:89–94. https://doi.org/10.1016/j.cplett.2017.07.042.

44. St-Amand L, Gagnon R, Packard TT, Savenkoff C. Effects of inorganic mercury on the respiration and the swimming activity of shrimp larvae, Pandalus borealis. *Comp Biochem Physiol - C Pharmacol Toxicol Endocrinol* 1999;122:33–43. https://doi.org/10.1016/S0742-8413(98)10071-3.

45. Sahu IP. Luminescence properties of cerium-doped di-strontium magnesium di-silicate phosphor by the solid-state reaction method. *Radiat Eff Defects Solids* 2016;171:544–64. https://doi.org/10.1080/10420150.2016.1217410.

46. Mao J, He X, Cao Z, Tang Y. Hydrogen titanates acting as inhibitor reservoirs and application in epoxy coatings. *Prog Org Coatings* 2019;127:394–400. https://doi.org/10.1016/j.porgcoat.2018.12.001.

47. Tuzen M, Karaman I, Citak D, Soylak M. Mercury(II) and methyl mercury determinations in water and fish samples by using solid phase extraction and cold vapour atomic absorption spectrometry combination. *Food Chem Toxicol* 2009;47:1648–52. https://doi.org/10.1016/j.fct.2009.04.024.

48. Gupta A, Vidyarthi SR, Sankararamakrishnan N. Enhanced sorption of mercury from compact fluorescent bulbs and contaminated water streams using functionalized multiwalled carbon nanotubes. *J Hazard Mater* 2014;274:132–44. https://doi.org/10.1016/j.jhazmat.2014.03.020.

49. Rogers PP, Jalal KF, Boyd JA. *An Introduction to Sustainable Development.* Earthscan; 2008. ISBN I 3 978-I-84407-521-8.

50. Game DN, Taide ST, Khan ZS, Ingale NB, Omanwar SK. Synthesis and Photoluminescence properties of $LiSrPO_4$: Eu^{2+} phosphor for Solid State Lighting. *AIP Conf Proc* 2016;020488:1–6. https://doi.org/10.1063/1.4946539.

51. Gomes IS, de Carvalho DC, Oliveira AC, Rodríguez-Castellón E, Tehuacanero-Cuapa S, Freire PTC, et al. On the reasons for deactivation of titanate nanotubes with metals catalysts in the acetalization of glycerol with acetone. *Chem Eng J* 2018;334:1927–42. https://doi.org/10.1016/j.cej.2017.11.112.

52. Snow DR. Microchronology and demographic evidence relating to the size of pre-columbian north american Indian populations. *Science* 1995;268:1601–4. https://doi.org/10.1126/science.268.5217.1601.

53. Lu X, Hoang S, Tang W, Du S, Wang S, Liu F, et al. Direct synthesis of conformal layered protonated titanate nanoarray coatings on various substrate surfaces boosted by low-temperature microwave-assisted hydrothermal synthesis. *ACS Appl Mater Interfaces* 2018;10:35164–74. https://doi.org/10.1021/acsami.8b11801.

54. Katogi A, Kubota K, Chihara K, Miyamoto K, Hasegawa T, Komaba S. Synthesis and electrochemical performance of C-base-centered lepidocrocite-like titanates for Na-ion batteries. *ACS Appl Energy Mater* 2018;1:3630–5. https://doi.org/10.1021/acsaem.8b00345.

55. Yuan H, Ma S, Wang X, Long H, Zhao X, Yang D, et al. Ultra-high adsorption of cationic methylene blue on two dimensional titanate nanosheets. *RSC Adv* 2019;9:5891–4. https://doi.org/10.1039/c8ra10172h.

56. Kordás K, Mohl M, Kónya Z, Kukovecz Á. Layered titanate nanostructures: Perspectives for industrial exploitation. *Transl Mater Res* 2015;2:015003. https://doi.org/10.1088/2053-1613/2/1/015003.

57. Liu Y, Wei R, Ding W, Wang X, Song W, Sheng Z, et al. Porous Fe_3O_4 thin films by pulsed laser assisted chemical solution deposition at room temperature. *Appl Surf Sci* 2019;478:408–11. https://doi.org/10.1016/j.apsusc.2019.01.282.

58. Johnson NC, Manchester S, Sarin L, Gao Y, Kulaots I, Hurt RH. Mercury vapor release from broken compact fluorescent lamps and in situ capture by new nanomaterial sorbents. *Environ Sci Technol* 2008;42:5772–8. https://doi.org/10.1021/es8004392.

59. Chen M, Wang Z, Liu H, Wang X, Ma Y, Liu J. Synthesis of potassium magnesium titanate whiskers with high near-infrared reflectivity by the flux method. *Mater Lett* 2017;202:59–61. https://doi.org/10.1016/j.matlet.2017.05.072.

60. Zhu QQ, Wang L, Hirosaki N, Hao LY, Xu X, Xie RJ. Extra-broad band orange-emitting Ce^{3+}-Doped $Y_3Si_5N_9O$ phosphor for solid-state lighting: Electronic, crystal structures and luminescence properties. *Chem Mater* 2016;28:4829–39. https://doi.org/10.1021/acs.chemmater.6b02109.

61. Singh R, Dutta S. A review on H_2 production through photocatalytic reactions using TiO_2/TiO_2-assisted catalysts. *Fuel* 2018;220:607–20. https://doi.org/10.1016/j.fuel.2018.02.068.

62. Saravanan R, Manoj D, Qin J, Naushad M, Gracia F, Lee AF, et al. Mechanothermal synthesis of Ag/TiO_2 for photocatalytic methyl orange degradation and hydrogen production. *Process Saf Environ Prot* 2018;120:339–47. https://doi.org/10.1016/j.psep.2018.09.015.

63. Wang Z, Hu T, He H, Fu Y, Zhang X, Sun J, et al. Enhanced H_2 production of TiO_2/ZnO nanowires co-using solar and mechanical energy through piezo-photocatalytic effect. *ACS Sustain Chem Eng* 2018;6:10162–72. https://doi.org/10.1021/acssuschemeng.8b01480.

64. Pellegrino F, Sordello F, Minella M, Minero C, Maurino V. The role of surface texture on the photocatalytic H_2 production on TiO_2. *Catalysts* 2019;9:32. https://doi.org/10.3390/catal9010032.

65. Hosseini MS, Belador F. Investigation on structural, optical, and magnetic properties of Dy-doped-CeO_2 nanoparticles synthesized by microwave-induced combustion method. *Synth React Inorganic, Met Nano-Metal Chem* 2016;46:950–7. https://doi.org/10.1080/15533174.2015.1004418.

66. Harabi A, Kasrani S, Foughali L, Serradj I, Benhassine MT, Kitouni S. Effect of TiO_2 additions on densification and mechanical properties of new mulltifunction resistant porcelains using economic raw materials. *Ceram Int* 2017;43:5547–56. https://doi.org/10.1016/j.ceramint.2017.01.081.

67. Lee EJ, An AK, Hadi P, Lee S, Woo YC, Shon HK. Advanced multi-nozzle electrospun functionalized titanium dioxide/polyvinylidene fluoride-co-hexafluoropropylene (TiO_2/PVDF-HFP) composite membranes for direct contact membrane distillation. *J Memb Sci* 2017;524:712–20. https://doi.org/10.1016/j.memsci.2016.11.069.

68. Zhang S, Peng LM, Chen Q, Du GH, Dawson G, Zhou WZ. Formation mechanism of [formula presented] nanotubes. *Phys Rev Lett* 2003;91:2–5. https://doi.org/10.1103/PhysRevLett.91.256103.

69. Yang Y, Li H, Zhao H, Qu R, Zhang S, Hu W, et al. Structure and crystal phase transition effect of Sn doping on anatase TiO_2 for dichloromethane decomposition. *J Hazard Mater* 2019;371:156–64. https://doi.org/10.1016/j.jhazmat.2019.02.103.

70. Tournoux M, Marchand R, Brohan L. Layered $K_2Ti_4O_9$ and the open metastable TiO_2(B) structure. *Prog Solid State Chem* 1986;17:33–52. https://doi.org/10.1016/0079-6786(86)90003-8.

71. Tudela RG, Gonçalves KA, Rocca RR. Author's accepted manuscript. *Radiat Phys Chem* 2018. https://doi.org/10.1016/j.radphyschem.2018.12.013.

72. Luo L, Xia L, Tan W, Li J, Barrow CJ, Yang W, et al. The TiO_2 (B) nano-belts with excellent performance prepared via alkaline stirring hydrothermal method and its application to remove 17α-ethynylestradiol. *Environ Sci Pollut Res* 2018;2. https://doi.org/10.1007/s11356-018-3122-8.

73. Nowak M, Sans-Merce M, Lemesre C, Elmiger R, Damet J. Eye lens monitoring programme for medical staff involved in fluoroscopy guided interventional procedures in Switzerland. *Phys Medica* 2019;57:33–40. https://doi.org/10.1016/j.ejmp.2018.12.001.

74. Maheu C, Cardenas L, Puzenat E, Afanasiev P, Geantet C. UPS and UV spectroscopies combined to position the energy levels of TiO_2 anatase and rutile nanopowders. *Phys Chem Chem Phys* 2018;20:25629–37. https://doi.org/10.1039/c8cp04614j.

75. Manjceevan A, Bandara J. Systematic stacking of PbS/CdS/CdSe multi-layered quantum dots for the enhancement of solar cell efficiency by harvesting wide solar spectrum. *Electrochim Acta* 2018;271:567–75. https://doi.org/10.1016/j.electacta.2018.03.193.

76. World Population estimates by the US Census Bureau. USCB.

77. Suzuki S, Kozawa T, Murakami T, Naito M. Mechanochemical-hydrothermal synthesis of layered lithium titanate hydrate nanotubes at room temperature and their conversion to $Li_4Ti_5O_{12}$. *Mater Res Bull* 2017;90:218–23. https://doi.org/10.1016/j.materresbull.2017.02.011.

78. He G, Zhang J, Hu Y, Bai Z, Wei C. Dual-template synthesis of mesoporous TiO_2 nanotubes with structure-enhanced functional photocatalytic performance. *Appl Catal B Environ* 2019:301–12. https://doi.org/10.1016/j.apcatb.2019.03.027.

79. Bai Q, Lavenas M, Vauriot L, Le Tréquesser Q, Hao J, Weill F, et al. Hydrothermal transformation of titanate scrolled nanosheets to anatase over a wide pH range and contribution of triethanolamine and oleic acid to control the morphology. *Inorg Chem* 2019;58:2588–98. https://doi.org/10.1021/acs.inorgchem.8b03197.

80. Preda S, Anastasescu C, Balint I, Umek P, Sluban M, Negrila CC, et al. Charge separation and ROS generation on tubular sodium titanates exposed to simulated solar light. *Appl Surf Sci* 2019;470:1053–63. https://doi.org/10.1016/j.apsusc.2018.11.194.

81. Maluangnont T, Wuttitham B, Hongklai P, Khunmee P, Tippayasukho S, Chanlek N, et al. An unusually acidic and thermally stable cesium titanate $Cs_xTi_{2-y}M_yO_4$ (x=0.67 or 0.70; M=vacancy or Zn). *Inorg Chem* 2019;58:6885–92. https://doi.org/10.1021/acs.inorgchem.9b00369.

82. Abass SAH, Seriani N. Structural and electronic properties of $Na_2Ti_3O_7$ and $H_2Ti_3O_7$. *Phys Status Solidi Basic Res* 2018;255:1–6. https://doi.org/10.1002/pssb.201700612.

83. Hu D, Miao L, Zhang Z, Li L, Wang Y, Cheng H, et al. One-dimensional piezoelectric $BaTiO_3$ polycrystal of topochemical mesocrystal conversion from layered $H_2Ti_4O_9$ H_2O single crystal. *Cryst. Growth Des* 2018;18:7264–74. https://doi.org/10.1021/acs.cgd.8b00351.

84. Islam MT, Sharmin N, Rance GA, Titman JJ, Parsons AJ, Hossain KMZ, et al. The effect of MgO/TiO_2 on structural and crystallization behavior of near invert phosphate-based glasses. *J Biomed Mater Res - Part B Appl Biomater* 2019;3:1–13. https://doi.org/10.1002/jbm.b.34421.

85. Thatcher A, Nayak R, Waterson P. Human factors and ergonomics systems-based tools for understanding and addressing global problems of the twenty-first century. *Ergonomics* 2019;0:1–41. https://doi.org/10.1080/00140139.2019.1646925.

86. Haut R, Wilcox A, Williams T, Burnett D. Overview of industry advancements in environmental awareness. *Proc - SPE Annu Tech Conf Exhib* 2018;2018:1–32.

87. Schmuch R, Wagner R, Hörpel G, Placke T, Winter M. Performance and cost of materials for lithium-based rechargeable automotive batteries. *Nat Energy* 2018;3:267–78. https://doi.org/10.1038/s41560-018-0107-2.

88. Lu J, Chen Z, Pan F, Cui Y, Amine K. High-performance anode materials for rechargeable lithium-ion batteries. *Electrochem Energy Rev* 2018;1:35–53. https://doi.org/10.1007/s41918-018-0001-4.

89. Nasir S, Hussein MZ, Zainal Z, Yusof NA. Carbon-based nanomaterials/allotropes: A glimpse of their synthesis, properties and some applications. *Materials (Basel)* 2018;11:1–24. https://doi.org/10.3390/ma11020295.

90. Liu J, Wang J, Xu C, Jiang H, Li C, Zhang L, et al. Advanced energy storage devices: Basic principles, analytical methods, and rational materials design. *Adv Sci* 2018;5. https://doi.org/10.1002/advs.201700322.

91. Cook JB, Lin TC, Kim HS, Siordia A, Dunn BS, Tolbert SH. Suppression of electrochemically driven phase transitions in nanostructured MoS_2 pseudoca-pacitors probed using operando X-ray diffraction. *ACS Nano* 2019;13:1223–31. https://doi.org/10.1021/acsnano.8b06381.

92. Deng J, Gong Q, Ye H, Feng K, Zhou J, Zha C, et al. Rational synthesis and assembly of Ni_3S_4 nanorods for enhanced electrochemical sodium-ion storage. *ACS Nano* 2018;12:1829–36. https://doi.org/10.1021/acsnano.7b08625.

93. Newell P, Lane R. A climate for change? The impacts of climate change on energy politics. *Cambridge Rev Int Aff* 2018;33(3):347–64. https://doi.org/10.1080/09557571.2018.1508203.

94. Nong D. General equilibrium economy-wide impacts of the increased energy taxes in Vietnam. *Energy Policy* 2018;123:471–81. https://doi.org/10.1016/j.enpol.2018.09.023.

95. Li J, Zhang D, Su B. The impact of social awareness and lifestyles on household carbon emissions in China. *Ecol Econ* 2019;160:145–55. https://doi.org/10.1016/j.ecolecon.2019.02.020.

96. Bekun FV, Alola AA, Sarkodie SA. Toward a sustainable environment: Nexus between CO_2 emissions, resource rent, renewable and nonrenewable energy in 16-EU countries. *Sci Total Environ* 2019;657:1023–9. https://doi.org/10.1016/j.scitotenv.2018.12.104.

97. Kahia M, Ben Jebli M, Belloumi M. Analysis of the impact of renewable energy consumption and economic growth on carbon dioxide emissions in 12 MENA countries. *Clean Technol Environ Policy* 2019;21:871–85. https://doi.org/10.1007/s10098-019-01676-2.

98. Miskinis V, Galinis A, Konstantinaviciute I, Lekavicius V, Neniskis E. Comparative analysis of the energy sector development trends and forecast of final energy demand in the Baltic States. *Sustainability* 2019;11. https://doi.org/10.3390/su11020521.

99. Clauser C, Ewert M. The renewables cost challenge: Levelized cost of geother-mal electric energy compared to other sources of primary energy – Review and case study. *Renew Sustain Energy Rev* 2018;82:3683–93. https://doi.org/10.1016/j.rser.2017.10.095.

100. Sahu BK. Wind energy developments and policies in China: A short review. *Renew Sustain Energy Rev* 2018;81:1393–405. https://doi.org/10.1016/j.rser.2017.05.183.

101. Uddin MN, Taweekun J, Techato K, Rahman MA, Mofijur M, Rasul MG. Sustainable biomass as an alternative energy source: Bangladesh perspec-tive. *Energy Procedia* 2019;160:648–54. https://doi.org/10.1016/j.egypro.2019.02.217.

102. Karatayev M, Clarke ML. A review of current energy systems and green energy potential in Kazakhstan. *Renew Sustain Energy Rev* 2016;55:491–504. https://doi.org/10.1016/j.rser.2015.10.078.

103. Hosseini SE, Wahid MA. Hydrogen production from renewable and sus-tainable energy resources: Promising green energy carrier for clean devel-opment. *Renew Sustain Energy Rev* 2016;57:850–66. https://doi.org/10.1016/j.rser.2015.12.112.

104. Olanrewaju RM, Olatunji OW, Akpan GP. Impacts of climate variability on hydroelectric power generation in Shiroro station, Nigeria. *Iran J Energy Environ* 2018;9:197–203.
105. Ogungbemi E, Wilberforce T, Alanzi A, Ijaodola O, Khatib FN, Hassan Z El, et al. The Future of Renewable Energy, Barriers and Solution. 11th International Conference on Sustainable Energy and Environmental Protection, Glasgow, Scotland, UK, 2018.
106. Moharramian A, Soltani S, Rosen MA, Mahmoudi SMS, Jafari M. Conventional and enhanced thermodynamic and exergoeconomic analyses of a photovoltaic combined cycle with biomass post firing and hydrogen production. *Appl Therm Eng* 2019;160:113996. https://doi.org/10.1016/j.applthermaleng.2019.113996.
107. Daryono D, Wahyud S, Suharnomo S. The development of green energy policy planning model to improve economic growth in Indonesia. *Int J Energy Econ Policy* 2019;9:216–23. https://doi.org/10.32479/ijeep.7779.
108. Omer AM. Energy, environment and sustainable development. *Renew Sustain Energy Rev* 2008;12:2265–300. https://doi.org/10.1016/j.rser.2007.05.001.
109. Høyer KG. The history of alternative fuels in transportation: The case of electric and hybrid cars. *Util Policy* 2008;16:63–71. https://doi.org/10.1016/j.jup.2007.11.001.
110. Fathabadi H. Utilizing solar and wind energy in plug-in hybrid electric vehicles. *Energy Convers Manag* 2018;156:317–28. https://doi.org/10.1016/j.enconman.2017.11.015.
111. Momirlan M, Veziroglu TN. The properties of hydrogen as fuel tomorrow in sustainable energy system for a cleaner planet. *Int J Hydrogen Energy* 2005;30:795–802. https://doi.org/10.1016/j.ijhydene.2004.10.011.
112. Marzi G, Caputo A, Garces E, Dabic M. A three decade mixed-method bibliometric investigation of the IEEE transactions on engineering management. *IEEE Trans Eng Manag* 2018:1–14. https://doi.org/10.1109/TEM.2018.2870648.
113. Liu Z, Liu Y, He BJ, Xu W, Jin G, Zhang X. Application and suitability analysis of the key technologies in nearly zero energy buildings in China. *Renew Sustain Energy Rev* 2019;101:329–45. https://doi.org/10.1016/j.rser.2018.11.023.
114. Li C, Meckler SM, Smith ZP, Bachman JE, Maserati L, Long JR, et al. Engineered transport in microporous materials and membranes for clean energy technologies. *Adv Mater* 2018;30. https://doi.org/10.1002/adma.201704953.
115. Nikolaidis P, Poullikkas A. Cost metrics of electrical energy storage technologies in potential power system operations. *Sustain Energy Technol Assess* 2018;25:43–59. https://doi.org/10.1016/j.seta.2017.12.001.
116. Poizot P, Dolhem F. Clean energy new deal for a sustainable world: From non-CO_2 generating energy sources to greener electrochemical storage devices. *Energy Environ Sci* 2011;4:2003–19. https://doi.org/10.1039/c0ee00731e.
117. Mozetič M, Vesel A, Primc G, Eisenmenger-Sittner C, Bauer J, Eder A, et al. Recent developments in surface science and engineering, thin films, nanoscience, biomaterials, plasma science, and vacuum technology. *Thin Solid Films* 2018;660:120–60. https://doi.org/10.1016/j.tsf.2018.05.046.
118. Abdalla AM, Hossain S, Azad AT, Petra PMI, Begum F, Eriksson SG, et al. Nanomaterials for solid oxide fuel cells: A review. *Renew Sustain Energy Rev* 2018;82:353–68. https://doi.org/10.1016/j.rser.2017.09.046.
119. Boldrin P, Brandon NP. Progress and outlook for solid oxide fuel cells for transportation applications. *Nat Catal* 2019;2:571–7. https://doi.org/10.1038/s41929-019-0310-y.

120. Leal-arcas R. *Climate and Energy Protection in the EU and China.* Springer International Publishing; 2019. https://doi.org/10.1007/978-3-319-99837-4.
121. Zhang Y, Zhang Q, Pan B. Impact of affluence and fossil energy on China carbon emissions using STIRPAT model. *Environ Sci Pollut Res* 2019;26:18814–24. https://doi.org/10.1007/s11356-019-04950-4.
122. Bazán J, Rieradevall J, Gabarrell X, Vázquez-Rowe I. Low-carbon electricity production through the implementation of photovoltaic panels in rooftops in urban environments: A case study for three cities in Peru. *Sci Total Environ* 2018;622–623:1448–62. https://doi.org/10.1016/j.scitotenv.2017.12.003.
123. Abid M, Abid Z, Sagin J, Murtaza R, Sarbassov D, Shabbir M. Prospects of floating photovoltaic technology and its implementation in Central and South Asian Countries. *Int J Environ Sci Technol* 2019;16:1755–62. https://doi.org/10.1007/s13762-018-2080-5.
124. Santhoshkumar A, Muthu Dinesh Kumar R, Babu D, Thangarasu V, Anand R. Effective utilization of high-grade energy through thermochemical conversion of different wastes. *IEEE Trans Eng Manag* 2019;67:4–17. https://doi.org/10.1007/978-981-13-3281-4_11.
125. Gottesfeld S, Dekel DR, Page M, Bae C, Yan Y, Zelenay P, et al. Anion exchange membrane fuel cells: Current status and remaining challenges. *J Power Sources* 2018;375:170–84. https://doi.org/10.1016/j.jpowsour.2017.08.010.
126. Venâncio SA, Moreira Sarruf BJ, Gomes GG, Valadão de Miranda PE. Multifunctional macroporous solid oxide fuel cell anode with active nanosized ceramic electrocatalyst. *Int J Hydrogen Energy* 2019. https://doi.org/10.1016/j.ijhydene.2019.06.006.
127. Hou J, Shao Y, Ellis MW, Moore RB, Yi B. Graphene-based electrochemical energy conversion and storage: Fuel cells, supercapacitors and lithium ion batteries. *Phys Chem Chem Phys* 2011;13:15384–402. https://doi.org/10.1039/c1cp21915d.
128. Nacimiento F, Cabello M, Alcántara R, Pérez-Vicente C, Lavela P, Tirado JL. Exploring an aluminum ion battery based on molybdite as working electrode and ionic liquid as electrolyte. *J Electrochem Soc* 2018;165:A2994–9. https://doi.org/10.1149/2.0391813jes.
129. Yildiz O, Dirican M, Fang X, Fu K, Jia H, Stano K, et al. Hybrid carbon nanotube fabrics with sacrificial nanofibers for flexible high performance lithium-ion battery anodes. *J Electrochem Soc* 2019;166:A473–9. https://doi.org/10.1149/2.0821902jes.
130. Schwenke KU, Solchenbach S, Demeaux J, Lucht BL, Gasteiger HA. The impact of CO_2 evolved from VC and FEC during formation of graphite anodes in lithium-ion batteries. *J Electrochem Soc* 2019;166:A2035–47. https://doi.org/10.1149/2.0821910jes.
131. Naghdi S, Rhee KY, Hui D, Park SJ. A review of conductive metal nanomaterials as conductive, transparent, and flexible coatings, thin films, and conductive fillers: Different deposition methods and applications. *Coatings* 2018;8. https://doi.org/10.3390/coatings8080278.
132. Singh R. Energy sufficiency aspirations of India and the role of renewable resources: Scenarios for future. *Renew Sustain Energy Rev* 2018;81:2783–95. https://doi.org/10.1016/j.rser.2017.06.083.
133. Cheng H, He X, Fan Z, Ouyang J. Flexible quasi-solid state ionogels with remarkable seebeck coefficient and high thermoelectric properties. *Adv Energy Mater* 2019;9:1901085. https://doi.org/10.1002/aenm.201901085.

134. Bowen CR, Kim HA, Weaver PM, Dunn S. Piezoelectric and ferroelectric materials and structures for energy harvesting applications. *Energy Environ Sci* 2014;7:25–44. https://doi.org/10.1039/c3ee42454e.
135. Russ B, Glaudell A, Urban JJ, Chabinyc ML, Segalman RA. Organic thermoelectric materials for energy harvesting and temperature control. *Nat Rev Mater* 2016;1. https://doi.org/10.1038/natrevmats.2016.50.
136. Kim CS, Yang HM, Lee J, Lee GS, Choi H, Kim YJ, et al. Self-powered wearable electrocardiography using a wearable thermoelectric power generator. *ACS Energy Lett* 2018;3:501–7. https://doi.org/10.1021/acsenergylett.7b01237.
137. Thielen M, Sigrist L, Magno M, Hierold C, Benini L. Human body heat for powering wearable devices: From thermal energy to application. *Energy Convers Manag* 2017;131:44–54. https://doi.org/10.1016/j.enconman.2016.11.005.
138. Shirakawa H, Louis EJ, MacDiarmid AG, Chiang CK, Heeger AJ. Synthesis of electrically conducting organic polymers: Halogen derivatives of polyacetylene, (CH)x. *J Chem Soc Chem Commun* 1977:578–80. https://doi.org/10.1039/C39770000578.
139. Wu D, Zhao LD, Hao S, Jiang Q, Zheng F, Doak JW, et al. Origin of the high performance in GeTe-based thermoelectric materials upon Bi_2Te_3 doping. *J Am Chem Soc* 2014;136:11412–9. https://doi.org/10.1021/ja504896a.
140. Garoboldi E, Colombo L, Fagiani D, Li Z, Methods to characterize effective thermal conductivity, diffusivity and thermal response in change materials. *Materials* 2019;12(16): 2552.
141. Bedenko S, Karengin A, Ghal-Eh N, Alekseev N, Knyshev V, Shamanin I. Thermo-physical properties of dispersion nuclear fuel for a new-generation reactors: A computational approach. *AIP Conf Proc* 2019;2101. https://doi.org/10.1063/1.5099594.
142. Zhang Y, Heo YJ, Park M, Park SJ. Recent advances in organic thermoelectric materials: Principle mechanisms and emerging carbon-based green energy materials. *Polymers (Basel)* 2019;11. https://doi.org/10.3390/polym11010167.
143. Oztan C, Ballikaya S, Ozgun U, Karkkainen R, Celik E. Additive manufacturing of thermoelectric materials via fused filament fabrication. *Appl Mater Today* 2019;15:77–82. https://doi.org/10.1016/j.apmt.2019.01.001.
144. Jang E, Poosapati A, Jang N, Hu L, Duffy M, Zupan M, et al. Thermoelectric properties enhancement of p-type composite films using wood-based binder and mechanical pressing. *Sci Rep* 2019;9:1–10. https://doi.org/10.1038/s41598-019-44225-z.
145. Boudouris BW, Yee S. Structure, properties and applications of thermoelectric polymers. *J Appl Polym Sci* 2017;134. https://doi.org/10.1002/app.44456.
146. Gao X, He M, Liu B, Hu J, Wang Y, Liang Z, et al. Topological Design of Inorganic-Organic Thermoelectric Nanocomposites Based on "electron-Percolation Phonon-Insulator" Concept. *ACS Appl Energy Mater* 2018;1:2927–33. https://doi.org/10.1021/acsaem.8b00615.
147. Wan C, Gu X, Dang F, Itoh T, Wang Y, Sasaki H, et al. Flexible n-type thermoelectric materials by organic intercalation of layered transition metal dichalcogenide TiS_2. *Nat Mater* 2015;14:622–7. https://doi.org/10.1038/nmat4251.
148. Wan C, Tian R, Kondou M, Yang R, Zong P, Koumoto K. Ultrahigh thermoelectric power factor in flexible hybrid inorganic-organic superlattice. *Nat Commun* 2017;8. https://doi.org/10.1038/s41467-017-01149-4.

149. Wan C, Kodama Y, Kondo M, Sasai R, Qian X, Gu X, et al. Dielectric mismatch mediates carrier mobility in organic-intercalated layered TiS_2. *Nano Lett* 2015;15:6302–8. https://doi.org/10.1021/acs.nanolett.5b01013.
150. Shin S, Roh JW, Kim HS, Chen R. Role of surfactant on thermoelectric behaviors of organic-inorganic composites. *J Appl Phys* 2018;123. https://doi.org/10.1063/1.5033920.
151. Malen JA, Yee SK, Majumdar A, Segalman RA. Fundamentals of energy transport, energy conversion, and thermal properties in organic-inorganic heterojunctions. *Chem Phys Lett* 2010;491:109–22. https://doi.org/10.1016/j.cplett.2010.03.028.
152. Pietrak K, Wiśniewski T. A review of models for effective thermal conductivity of composite materials. *J Power Technol* 2015;95:14–24.
153. Koga T, Cronin SB, Dresselhaus MS, Liu JL, Wang KL. Experimental proof-of-principle investigation of enhanced Z_3DT in (001) oriented Si/Ge superlattices. *Appl Phys Lett* 2000;77:1490–2. https://doi.org/10.1063/1.1308271.
154. Caylor JC, Coonley K, Stuart J, Colpitts T, Venkatasubramanian R. Enhanced thermoelectric performance in PbTe-based superlattice structures from reduction of lattice thermal conductivity. *Appl Phys Lett* 2005;87:1–4. https://doi.org/10.1063/1.1992662.
155. Ohta H, Kim S, Mune Y, Mizoguchi T, Nomura K, Ohta S, et al. Giant thermoelectric Seebeck coefficient of a two-dimensional electron gas in $SrTiO_3$. *Nat Mater* 2007;6:129–34. https://doi.org/10.1038/nmat1821.
156. Tian Y, Sakr MR, Kinder JM, Liang D, MacDonald MJ, Qiu RLJ, et al. One-dimensional quantum confinement effect modulated thermoelectric properties in InAs nanowires. *Nano Lett* 2012;12:6492–7. https://doi.org/10.1021/nl304194c.
157. Petsagkourakis I, Tybrandt K, Crispin X, Ohkubo I, Satoh N, Mori T. Thermoelectric materials and applications for energy harvesting power generation. *Sci Technol Adv Mater* 2018;19:836–62. https://doi.org/10.1080/14686996.2018.1530938.
158. Takashiri M, Miyazaki K, Tanaka S, Kurosaki J, Nagai D, Tsukamoto H. Effect of grain size on thermoelectric properties of n -type nanocrystalline bismuth-telluride based thin films. *J Appl Phys* 2008;104. https://doi.org/10.1063/1.2990774.
159. Varghese T, Hollar C, Richardson J, Kempf N, Han C, Gamarachchi P, et al. High-performance and flexible thermoelectric films by screen printing solution-processed nanoplate crystals. *Sci Rep* 2016;6:6–11. https://doi.org/10.1038/srep33135.
160. Gutierrez A, Miró L, Gil A, Rodríguez-Aseguinolaza J, Barreneche C, Calvet N, et al. Advances in the valorization of waste and by-product materials as thermal energy storage (TES) materials. *Renew Sustain Energy Rev* 2016;59:763–83. https://doi.org/10.1016/j.rser.2015.12.071.
161. Hassan A, Laghari MS, Rashid Y. Micro-encapsulated phase change materials: A review of encapsulation, safety and thermal characteristics. *Sustain* 2016;8. https://doi.org/10.3390/su8101046.
162. Park CS, Hong MH, Cho HH, Park HH. Enhancement of Seebeck coefficient of mesoporous $SrTiO_3$ with V-group elements V, Nb, and Ta substituted for Ti. *J Eur Ceram Soc* 2018;38:125–30. https://doi.org/10.1016/j.jeurceramsoc.2017.08.021.
163. Vidyasagar CC, Muñoz Flores BM, Jiménez Pérez VM. Recent advances in synthesis and properties of hybrid halide perovskites for photovoltaics. *Nano-Micro Lett* 2018;10. https://doi.org/10.1007/s40820-018-0221-5.

164. Xiao X, Widenmeyer M, Mueller K, Scavini M, Checchia S, Castellano C, et al. A squeeze on the perovskite structure improves the thermoelectric performance of Europium Calcium Titanates. *Mater Today Phys* 2018;7:96–105. https://doi.org/10.1016/j.mtphys.2018.11.009.

165. Kadam AR, Dhoble SJ. Synthesis and luminescence study of Eu^{3+}-doped $SrYAl_3O_7$ phosphor. *Luminescence* 2019. https://doi.org/10.1002/bio.3681.

166. Kadam AR, Nair GB, Dhoble SJ. Insights into the extraction of mercury from fluorescent lamps: A review. *J Environ Chem Eng* 2019;7:103279. https://doi.org/10.1016/j.jece.2019.103279.

167. Li Z, Zhong B, Cao Y, Zhang S, Lv Y, Mu Z, et al. Energy transfer and tunable luminescence properties in $Y_3Al_2Ga_3O_{12} : Tb^{3+}$, Eu^{3+} phosphors. *J Alloys Compd* 2019;787:672–82. https://doi.org/10.1016/j.jallcom.2019.02.154.

168. Hassan Z, Abd HR, Alsultany FH, Omar AF, Ahmed NM. Investigation of sintering temperature and Ce^{3+} concentration in YAG: Ce phosphor powder prepared by microwave combustion for white-light-emitting diode luminance applications. *Mater Chem Phys* 2019;229:22–31. https://doi.org/10.1016/j.matchemphys.2019.02.090.

169. Gu G, Xiang W, Yang C, Liang X. Synthesis and luminescence properties of a H_2 annealed Mn-doped $Y_3Al_5O_{12}$:Ce^{3+} single crystal for WLEDs. *CrystEngComm* 2015;17:4554–61. https://doi.org/10.1039/c5ce00641d.

170. Du P, Huang X, Yu JS. Facile synthesis of bifunctional Eu^{3+}-activated $NaBiF_4$ red-emitting nanoparticles for simultaneous white light-emitting diodes and field emission displays. *Chem Eng J* 2018;337:91–100. https://doi.org/10.1016/j.cej.2017.12.063.

171. Du P, Krishna Bharat L, Yu JS. Strong red emission in Eu^{3+}/Bi^{3+} ions codoped $CaWO_4$ phosphors for white light-emitting diode and field-emission display applications. *J Alloys Compd* 2015;633:37–41. https://doi.org/10.1016/j.jallcom.2015.01.287.

172. Basiricò L, Senanayak SP, Ciavatti A, Abdi-Jalebi M, Fraboni B, Sirringhaus H. Detection of X-rays by solution-processed cesium-containing mixed triple cation perovskite thin films. *Adv Funct Mater* 2019;1902346:1–9. https://doi.org/10.1002/adfm.201902346.

173. Jena AK, Kulkarni A, Miyasaka T. Halide perovskite photovoltaics: Background, status, and future prospects. *Chem Rev* 2019;119:3036–103. https://doi.org/10.1021/acs.chemrev.8b00539.

174. Kaabeche A, Bakelli Y. Renewable hybrid system size optimization considering various electrochemical energy storage technologies. *Energy Convers Manag* 2019;193:162–75. https://doi.org/10.1016/j.enconman.2019.04.064.

175. Gayen RN, Avvaru VS, Etacheri V. Carbon-based integrated devices for efficient photo-energy conversion and storage. In *Carbon Based Nanomaterials for Advanced thermal and Electrochemical Energy Storage and Conversion*, edited by Rajib Paul, Vinodkumar Etacheri, Yan Wang and Cheng-Te Lin. Elsevier Inc.; 2019. https://doi.org/10.1016/b978-0-12-814083-3.00014-7.

176. Pandey AK, Hossain MS, Tyagi V V., Abd Rahim N, Selvaraj JAL, Sari A. Novel approaches and recent developments on potential applications of phase change materials in solar energy. *Renew Sustain Energy Rev* 2018;82:281–323. https://doi.org/10.1016/j.rser.2017.09.043.

177. Bashir S, Hanumandla P, Huang HY, Liu JL. Nanostructured materials for advanced energy conversion and storage devices: Safety implications at end-of-life disposal. In *Nanostructured Materials for Next-Generation Energy Storage*

and Conversion: Fuel Cells, edited by Li Fan, Bashir S and Liu J. SpringerLink; 2018;4:517–42. https://doi.org/10.1007/978-3-662-56364-9_18.

178. Shukla AK, Prem Kumar T. Nanostructured electrode materials for electrochemical energy storage and conversion. *Wiley Interdiscip Rev Energy Environ* 2013;2:14–30. https://doi.org/10.1002/wene.48.

179. Xing C, Isaacowitz DM. Master En Ingenieria Ambiental 2006–07. *Motiv Emot* 2006:2878–87. https://doi.org/10.1002/adma.200800627.

180. Bruce PG, Freunberger SA, Hardwick LJ, Tarascon JM. Li_gO_2 and Li_gS batteries with high energy storage. *Nat Mater* 2012;11:19–29. https://doi.org/10.1038/nmat3191.

181. Cheng Z, Xiao Z, Pan H, Wang S, Wang R. Elastic sandwich-type rGO–VS2/S composites with high tap density: Structural and chemical cooperativity enabling lithium–sulfur batteries with high energy density. *Adv Energy Mater* 2018;8:1–12. https://doi.org/10.1002/aenm.201702337.

182. Wang B, Ryu J, Choi S, Song G, Hong D, Hwang C, et al. Folding graphene film yields high areal energy storage in lithium-ion batteries. *ACS Nano* 2018;12:1739–46. https://doi.org/10.1021/acsnano.7b08489.

183. Li Q, Wang Q. Ferroelectric polymers and their energy-related applications. *Macromol Chem Phys* 2016;217:1228–44. https://doi.org/10.1002/macp.201500503.

184. Setter N. What is a ferroelectric–a materials designer perspective. *Ferroelectrics* 2016;500:164–82. https://doi.org/10.1080/00150193.2016.1232104.

185. Narita F, Fox M. A review on piezoelectric, magnetostrictive, and magnetoelectric materials and device technologies for energy harvesting applications. *Adv Eng Mater* 2018;20:1–22. https://doi.org/10.1002/adem.201700743.

186. Stadlober B, Zirkl M, Irimia-Vladu M. Route towards sustainable smart sensors: Ferroelectric polyvinylidene fluoride-based materials and their integration in flexible electronics. *Chem Soc Rev* 2019;48:1787–825. https://doi.org/10.1039/c8cs00928g.

187. Zhang G, Li M, Li H, Wang Q, Jiang S. Harvesting energy from human activity: Ferroelectric energy harvesters for portable, implantable, and biomedical Electronics. *Energy Technol* 2018;6:791–812. https://doi.org/10.1002/ente.201700622.

188. Kishore RA, Priya S. A review on low-grade thermal energy harvesting: Materials, methods and devices. *Materials (Basel)* 2018;11. https://doi.org/10.3390/ma11081433.

189. Li Q, Yao F-Z, Liu Y, Zhang G, Wang H, Wang Q. High-temperature dielectric materials for electrical energy storage. *Annu Rev Mater Res* 2018;48:219–43. https://doi.org/10.1146/annurev-matsci-070317-124435.

190. Salcedo S, Nieto A, Vallet-Regi M. Hydroxiapatite/β tricalcium phosphate/agarose Macroporous scaffolds for bone tissue engineering. *Chem Eng J* 2008;137:62–71. https://doi.org/10.1016/j.cej.2007.09.011.

191. Reynolds S, Welter K, Smirnov V. Silicon thin films: Functional materials for energy, healthcare, and IT applications. *Phys Status Solidi Appl Mater Sci* 2019;216:1–12. https://doi.org/10.1002/pssa.201800847.

192. Ofoegbuna T, Darapaneni P, Sahu S, Plaisance C, Dorman JA. Stabilizing the B-site oxidation state in ABO_3 perovskite nanoparticles. *Nanoscale* 2019;11:14303–11. https://doi.org/10.1039/c9nr04155a.

193. Sando D, Yang Y, Paillard C, Dkhil B, Bellaiche L, Nagarajan V. Epitaxial ferroelectric oxide thin films for optical applications. *Appl Phys Rev* 2018;5. https://doi.org/10.1063/1.5046559.

194. Rickman JM, Lookman T, Kalinin S V. Materials informatics: From the atomic-level to the continuum. *Acta Mater* 2019;168:473–510. https://doi.org/10.1016/j.actamat.2019.01.051.

195. Martin A, Khansur NH, Riess K, Webber KG. Frequency dependence of the relaxor-to-ferroelectric transition under applied electrical and mechanical fields. *J Eur Ceram Soc* 2019;39:1031–41. https://doi.org/10.1016/j.jeurceramsoc.2018.12.026.

196. Song H, Liu X, Wang B, Tang Z, Lu S. High production-yield solid-state carbon dots with tunable photoluminescence for white / multi-color light-emitting diodes. *Sci Bull* 2019. https://doi.org/10.1016/j.scib.2019.10.006.

197. Tung TX, Xu D, Zhang Y, Zhou Q, Wu Z. Removing humic acid from aqueous solution using titanium dioxide: A review. *Pol J Environ Stud* 2019;28:529–42. https://doi.org/10.15244/pjoes/85196.

198. Khan B, Raziq F, Faheem MB, Farooq MU, Hussain S, Ali F. Electronic and nanostructure engineering of bifunctional MoS_2 towards exceptional visible-light photocatalytic CO_2 reduction and pollutant degradation. *J Hazard Mater* 2020;381:120972. https://doi.org/10.1016/j.jhazmat.2019.120972.

199. Reddy KR, Jyothi MS, Raghu A V, Sadhu V, Naveen S, Aminabhavi TM. Nanocarbons-supported and polymers- supported titanium dioxide nanostructures as efficient photocatalysts for remediation of contaminated wastewater and hydrogen production. In *Nanophotocatalysis and Environmental Applications*, edited by Inamuddin, Asiri A and Lichtfouse E. SpringerLink Publications; 2020;139–169.

200. Bedia J, Rodriguez JJ, Belver C. C-modified TiO_2 using lignin as carbon precursor for the solar photocatalytic degradation of acetaminophen. *Chem Eng J* 2018. https://doi.org/10.1016/j.cej.2018.10.154.

201. Kusiak-Nejman E, Morawski AW. TiO_2/graphene-based nanocomposites for water treatment: A brief overview of charge carrier transfer, antimicrobial and photocatalytic performance. *Appl Catal B Environ* 2019;253:179–86. https://doi.org/10.1016/j.apcatb.2019.04.055.

202. Qian Y, Feng J, Wang H, Fan D, Jiang N, Wei Q, et al. Sandwich-type signal-off photoelectrochemical immunosensor based on dual suppression effect of PbS quantum dots / Co_3O_4 polyhedron as signal amplifi cation for procalcitonin detection. *Sensors Actuators B Chem* 2019;300:127001. https://doi.org/10.1016/j.snb.2019.127001.

203. Bhanvase BA, Shende TP, Sonawane SH. A review on graphene – TiO_2 and doped graphene – TiO_2 nanocomposite photocatalyst for water and wastewater treatment. *Environ Technol Rev* 2017;2515. https://doi.org/10.1080/21622515.2016.1264489.

204. Zangeneh H, Zinatizadeh AA, Feyzi M, Zinadini S, Bahnemann DW. PT SC. *Biochem Pharmacol* 2018. https://doi.org/10.1016/j.jece.2018.10.001.

205. Habchi R, Bechelany M. Recent progress on titanium dioxide nanomaterials for photocatalytic applications. *Chempluschem* 2018;11(18):3023–47. https://doi.org/10.1002/cssc.201800874.

206. Wang M, Ye M, Iocozzia J, Lin C, Lin Z. Plasmon-mediated solar energy conversion via photocatalysis in noble metal / semiconductor composites. *Adv Sci* 2016. https://doi.org/10.1002/advs.201600024.

207. Kuvarega AT, Mamba BB, Mamba BB. TiO_2-based photocatalysis: Toward visible light-responsive photocatalysts through doping and fabrication of

carbon-based nanocomposites. *Crit Rev Solid State Mater Sci* 2016;0:1–52. https://doi.org/10.1080/10408436.2016.1211507.

208. Etacheri V, Di C, Schneider J, Bahnemann D, Pillai SC. Visible-light activation of TiO_2 photocatalysts : Advances in theory and experiments. *J Photochem Photobiol C: Photochem Rev* 2015;25:1–29. https://doi.org/10.1016/j.jphotochemrev.2015.08.003.

209. Singh K, Harish S, Archana J, Navaneethan M, Shimomura M. Applied Surface Science Investigation of Gd-doped mesoporous TiO_2 spheres for environmental remediation and energy applications. *Appl Surf Sci* 2019;489:883–92. https://doi.org/10.1016/j.apsusc.2019.05.253.

210. Barawi M, Fresno F, Perez-ruiz R, de la Peña O'Shea VA. Photoelectrochemical hydrogen evolution driven by visible-to-UV photon upconversion. *ACS Appl Energy Mater* 2018;2:207–11. https://doi.org/10.1021/acsaem.8b01916.

211. Dhandayuthapani T, Sivakumar R, Ilangovan R, Gopalakrishnan C, Sanjeeviraja C, Sivanantharaja A. Efficient electrochromic performance of anatase TiO_2 thin films prepared by nebulized spray deposition method. *J Solid State Electrochem* 2018;22:1825–38.

212. Kumar P, Swart HC. Plasmonic metamaterial-based chemical converted graphene/TiO2/Ag thin films by a simple spray pyrolysis technique. *Phys B Phys Condens Matter* 2017;535:299–303. https://doi.org/10.1016/j.physb.2017.08.006.

213. Adyani SM, Ghorbani M. A comparative study of physicochemical and photocatalytic properties of visible light responsive Fe, Gd and P single and tri-doped TiO_2. *J Rare Earths* 2018;36:72–85. https://doi.org/10.1016/j.jre.2017.06.012.

214. Singh K, Harish S, Kristy AP, Shivani V, Archana J, Navaneethan M, Shimomura M, Hayakawa Y. Erbium doped TiO_2 interconnected mesoporous spheres as an efficient visible light catalyst for photocatalytic applications. *Appl Surf Sci* 2018;449:755–63. https://doi.org/10.1016/j.apsusc.2018.01.279.

215. Wang Y, Liu A, Ma D, Chen C. TiO_2 photocatalyzed C–H bond transformation for C–C coupling reactions. *Catalysts* 2018;8(9):355. https://doi.org/10.3390/catal8090355.

216. Mckenna B, Evans RC. Towards efficient spectral converters through materials design for luminescent solar devices. *Adv Mater* 2017;1606491:1–23. https://doi.org/10.1002/adma.201606491.

217. Tian J, Leng Y, Zhao Z, Xia Y. Carbon quantum dots/hydrogenated TiO_2 nanobelt heterostructures and their broad spectrum photocatalytic properties under UV, visible, and near-infrared irradiation. *Nano Energy* 2014:1–9. https://doi.org/10.1016/j.nanoen.2014.10.025.

218. Mahraz ZAS, Sahar MR, Ghoshal SK. Near-infrared up-conversion emission from erbium ions doped amorphous tellurite media: Judd-Ofelt evaluation. *J Alloys Compd* 2018;740:617–25. Elsevier B.V. https://doi.org/10.1016/j.jallcom.2017.12.314.

219. Zhang W, Yang S, Li J, Gao W, Deng Y, Dong W, et al. Visible-to-ultraviolet Upconvertion: Energy transfer, material matrix, and synthesis strategies. *Appl Catal B Environ* 2017;206:89–103. https://doi.org/10.1016/j.apcatb.2017.01.023.

220. Parnicka P, Grzyb T, Mikolajczyk A, Wang K, Kowalska E, Steinfeldt N, et al. Experimental and theoretical investigations of the influence of carbon on a Ho^{3+} -TiO_2 photocatalyst with Vis response. *J Colloid Interface Sci* 2019;549:212–24. https://doi.org/10.1016/j.jcis.2019.04.074.

221. León-Luis SF, Rodríguez-Mendoza UR, Lalla E, Lavín V. Temperature sensor based on the Er^{3+} green upconverted emission in a fluorotellurite glass. *Sens Actuators B Chem* 2011;158:208–13. https://doi.org/10.1016/j.snb.2011.06.005.

222. Tian Y, Xu R, Hu L, Zhang J. 2.7 μm fluorescence radiative dynamics and energy transfer between Er^{3+} and Tm^{3+} ions in fluoride glass under 800 nm and 980 nm excitation. *J Quant Spectrosc Radiat Transf* 2012;113:87–95. https://doi.org/10.1016/j.jqsrt.2011.09.016.

223. Zhang X, Daran E, Serrano C, Lahoz F. Up-conversion fluorescence in MBE-grown Nd^{3+}-doped LaF_3/CaF_2 waveguides. *J Lumin* 2000;89:1011–3.

224. Lv Y, Du W, Ren Y, Cai Z, Yu K, Zhang C, Chen Z, Wright DS. An integrated electrochromic supercapacitor based on nanostructured Er-containing titania using a Er(III)-doped polyoxotitanate cage. *Inorg Chem Front* 2016;3:1119–23 https://doi.org/10.1039/C6QI00114A.

225. Nicolás P, Caicedo A, Grzyb T, Mikolajczyk A, Roy JK, Wyrzykowska E, et al. Experimental and computational study of Tm-doped TiO_2: The e ff ect of Li^+ on Vis-response photocatalysis and luminescence. *Appl Catal B: Environ* 2019;252:138–51. https://doi.org/10.1016/j.apcatb.2019.03.051.

226. Lahiri R, Member S, Mondal A. Superior memory of Er doped TiO_2 nanowire MOS capacitor. *IEEE Electron Device Lett* 2018; 39(12):1856-9. https://doi.org/10.1109/LED.2018.2874272.

227. Van Hau D, Nhan DTT, Van Duc N, Tuyen VP, Nguyen TD, Hoa TT, Cuong ND. Structural design of near-infrared light-active $Cu/TiO_2/NaYF_4$: Yb, Er nanocomposite photocatalysts. *J Electron Mater* 2018;48:329–36. https://doi.org/10.1007/s11664-018-6717-4.

228. Cruz P, Hierro I, Pérez Y. Selective oxidation of thioanisole by titanium complexes immobilized on mesoporous silica nanoparticles : Elucidating the environment of titanium (iv) species. *Catal Sci Technol* 2018. https://doi.org/10.1039/c8cy01929k.

229. Kumada N, Imase A, Yanagida S, Takei T, Itoi N, Goto T. Hydrothermal synthesis of $KTi_2(PO_4)_3$, α-$Ti(HPO_4)_2 \cdot H_2O$ and γ-$Ti(PO_4)(H_2PO_4) \cdot 2H_2O$ from a lepidocrocite-type titanate. *J Asian Ceram Soc* 2019;7:361–67. https://doi.org/10.1080/21870764.2019.1649041.

230. Gu H, Wang J, Li Y, Wang Z, Fu Y. The core-shell-structured $NaYF_4:Er^{3+}$, Yb^{3+}@ $NaYF_4:Eu^{3+}$ nanocrystals as dual-mode and multifunctional luminescent mechanism for high-performance dye-sensitized solar cells. *Mater Res Bull* 2018;108:219–25. https://doi.org/10.1016/j.materresbull.2018.09.015.

231. Boltenkov IS, Kolobkova E V, Evstropiev SK. SC. Synthesis and characterization of transparent photocatalytic ZnO-Sm_2O_3 and ZnO-Er_2O_3 coatings. *Journal Photochem Photobiol A Chem* 2018. https://doi.org/10.1016/j.jphotochem.2018.09.016.

232. Tian C. Supramolecular self-assembly synthesis of ordered mesoporous TiO_2 from industrial $TiOSO_4$ solution and its photocatalytic activities. *Microporous Mesoporous Mater* 2019;286:176–81. https://doi.org/10.1016/j.micromeso.2019.05.047.

233. Kabir I, Sheppard L, Liu R, Yao Y, Zhu Q, et al.,Contamination of TiO_2 thin films-spin coated on rutile and fused silica substrates. *Surf Coat Technol* 2018;354:369–82. https://doi.org/10.1016/j.surfcoat.2018.09.009.

234. Han J, Hou X, Liu H, Li J, Yao J, Li D, et al. Photocurrent enhancement on TiO_2 nanotubes co- modified by N^+ implantation and combustion of graphene. *Mater Lett* 2018. https://doi.org/10.1016/j.matlet.2018.11.155.

235. Chouhan R, Baraskar P, Agrawal A, Gupta M, Sen P. Nonlinear optical responses of magnetron sputtered TiO_2 thin film. AIP Conference Proceedings, 2019;2115;030301.
236. Yoo J, Zazpe R, Cha G, Prikryl J, Hwang I, Macak JM. Electrochemistry communications uniform ALD deposition of Pt nanoparticles within 1D anodic TiO_2 nanotubes for photocatalytic H_2 generation. *Electrochem Commun* 2018;86:6–11. https://doi.org/10.1016/j.elecom.2017.10.017.
237. Giardina A, Tampieri F, Marotta E, Paradisi C. Air non-thermal plasma treatment of Irgarol 1051 deposited on TiO_2. *Chemosphere* 2018;210:653–61. https://doi.org/10.1016/j.chemosphere.2018.07.012.
238. Mazierski P, Lisowski W, Grzyb T, Winiarski MJ, Klimczuk T, Mikołajczyk A, et al. Enhanced photocatalytic properties of lanthanide-TiO_2 nanotubes: An experimental and theoretical study. *Applied Catal B, Environ* 2016. https://doi.org/10.1016/j.apcatb.2016.12.044.
239. Hussin SHA. Study the Nanoporous SiO_2 – TiO_2 Doped with Er and La Thin Films Properties A ntireflection and Self-Cleaning Applications. 2017. https://doi.org/10.13140/RG.2.2.23286.70729.
240. Suo H, Guo C, Li T. Broad-scope Thermometry Based on Dual-color Modulation Up-conversion Phosphor Ba5Gd8Zn4O21: Er3+/Yb3+. *J Phys Chem C* 2016; 120:2914–24. https://doi.org/10.1021/acs.jpcc.5b11786.
241. Liang L, Yulin Y, Mi Z, Ruiqing F, Lele Q, Xin W, et al. Enhanced performance of dye-sensitized solar cells based on TiO_2 with NIR-absorption and visible upconversion luminescence. *J Solid State Chem* 2013;198:459–65. https://doi.org/10.1016/j.jssc.2012.10.013.
242. Borlaf M, Caes S, Dewalque J, Teresa M, Moreno R, Cloots R, et al. Effect of the RE (RE=Eu, Er) doping on the structural and textural properties of mesoporous TiO_2 thin films obtained by evaporation induced self-assembly method. *Thin Solid Films* 2014;558:140–8. https://doi.org/10.1016/j.tsf.2014.03.002.
243. Obregón S, Kubacka A, Fernández-garcía M, Colón G. High-performance Er^{3+} – TiO_2 system: Dual up-conversion and electronic role of the lanthanide. *J Catal* 2013;299:298–306. https://doi.org/10.1016/j.jcat.2012.12.021.
244. Li Y, Wang Y, Kong J, Wang J. Synthesis and photocatalytic activity of TiO_2 nanotubes co-doped by erbium ions. *Appl Surf Sci* 2015;328:115–9. https://doi.org/10.1016/j.apsusc.2014.12.054.
245. Yang Y, Zhang C, Xu Y, Wang H, Li X, Wang C. Electrospun Er: TiO_2 nano fi brous films as efficient photocatalysts under solar simulated light. *Mater Lett* 2010;64:147–50. https://doi.org/10.1016/j.matlet.2009.10.028.
246. Er Y. Preparation and up-conversion fluorescence of rare earth. *J Solid State Chem* 2010;183:584–9. https://doi.org/10.1016/j.jssc.2010.01.004.
247. Falcomer D, Daldosso M, Cannas C, Musinu A, Lasio B, Enzo S, et al. A one-step solvothermal route for the synthesis of nanocrystalline anatase TiO_2 doped with lanthanide ions. *J Solid State Chem* 2006;179:2452–7. https://doi.org/10.1016/j.jssc.2006.04.043.
248. Obregón S, Colón G. Evidence of upconversion luminescence contribution to the improved photoactivity of erbium doped TiO_2 systems. *Chem Commun* 2012;48:7865–7. https://doi.org/10.1039/c2cc33391k.
249. Qin H, Shamso AE, Centeno A, Theodorou IG, Mihai AP, Ryan MP, et al. Enhancement of the upconversion photoluminescence of hexagonal phase $NaYF_4$:Yb^{3+}, Er^{3+} nanoparticles by mesoporous gold films. *Phys Chem Chem Phys* 2017;19:19159–67. https://doi.org/10.1039/c7cp01959a.

250. Castañeda-Contreras J, Marañón-Ruiz VF, Chiu-Zárate R, Pérez-Ladrón De Guevara H, Rodriguez R, Michel-Uribe C. Photocatalytic activity of erbium-doped TiO$_2$ nanoparticles immobilized in macro-porous silica films. *Mater Res Bull* 2012;47:290–5. https://doi.org/10.1016/j.materresbull.2011.11.021.

251. Lee DY, Kim JT, Park JH, Kim YH, Lee IK, Lee MH, et al. Effect of Er doping on optical band gap energy of TiO$_2$ thin films prepared by spin coating. *Curr Appl Phys* 2013;13:1301–5. https://doi.org/10.1016/j.cap.2013.03.025.

252. Zhao DD, Zeng A, Shang MS, Gao J. Long-term effects of recommendation on the evolution of online systems. *Chinese Phys Lett* 2013;30. https://doi.org/10.1088/0256-307X/30/11/118901.

253. Wu X, Yin S, Dong Q, Sato T. Blue/green/red colour emitting up-conversion phosphors coupled C-TiO$_2$ composites with UV, visible and NIR responsive photocatalytic performance. *Appl Catal B Environ* 2014;156–157:257–64. https://doi.org/10.1016/j.apcatb.2014.03.028.

254. Huang X, Jiang L, Li X, He A. Manipulating upconversion emission of cubic BaGdF5: Ce3+/Er3+/Yb3+ nanocrystals through controlling Ce3+ doping. *J Alloys Compd* 2017;721:374–82.

255. Obregón S, Kubacka A, Fernández-García M, Colón G. High-performance Er^{3+}-TiO$_2$ system: Dual up-conversion and electronic role of the lanthanide. *J Catal* 2013;299:298–306. https://doi.org/10.1016/j.jcat.2012.12.021.

256. Mao X, Yan B, Wang J, Shen J. Up-conversion fluorescence characteristics and mechanism of Er^{3+}-doped TiO$_2$ thin films. *Vacuum* 2014;102:38–42. https://doi.org/10.1016/j.vacuum.2013.10.026.

257. Bahtat A, Bouazaoui M, Bahtat M, Garapon C, Jacquier B, Mugnier J. Up-conversion fluorescence spectroscopy in Er^{3+}: TiO$_2$ planar waveguides prepared by a sol-gel process. *J Non Cryst Solids* 1996;202:16–22. https://doi.org/10.1016/0022-3093(96)00172-X.

258. Salhi R, Deschanvres JL. Efficient green and red up-conversion emissions in Er/Yb co-doped TiO$_2$ nanopowders prepared by hydrothermal-assisted sol–gel process. *J Lumin* 2016;176:250–9. https://doi.org/10.1016/j.jlumin.2016.03.011.

259. Khaki MRD, Sajjadi B, Raman AAA, Daud WMAW, Shmshirband S. Sensitivity analysis of the photoactivity of Cu-TiO$_2$/ZnO during advanced oxidation reaction by Adaptive Neuro-Fuzzy Selection Technique. *Meas J Int Meas Confed* 2016;77:155–74. https://doi.org/10.1016/j.measurement.2015.07.004.

260. Shang, J., Xu, W.W., Ye, C. et al. Synergistic effect of nitrate-doped TiO$_2$ aerosols on the fast photochemical oxidation of formaldehyde. *Sci Rep* 7, 1161 (2017). https://doi.org/10.1038/s41598-017-01396-x.

261. Reszczyńska J, Grzyb T, Wei Z, Klein M, Kowalska E, Ohtani B, et al. Photocatalytic activity and luminescence properties of RE^{3+}-TiO$_2$ nanocrystals prepared by sol-gel and hydrothermal methods. *Appl Catal B Environ* 2016;181:825–37. https://doi.org/10.1016/j.apcatb.2015.09.001.

262. Zheng Y, Wang W. Electrospunnanofibers of Er3+-doped TiO2 with photocatalyticactivity beyondtheabsorptionedge. *J Solid State Chem* 2014;210:206–12.

263. Liang CH, Hou MF, Zhou SG, Li FB, Liu CS, Liu TX, et al. The effect of erbium on the adsorption and photodegradation of orange I in aqueous Er^{3+}-TiO$_2$ suspension. *J Hazard Mater* 2006;138:471–8. https://doi.org/10.1016/j.jhazmat.2006.05.066.

264. Bhethanabotla VC, Russell DR, Kuhn JN. Assessment of mechanisms for enhanced performance of Yb/Er/titania photocatalysts for organic degradation: Role of rare earth elements in the titania phase. *Appl Catal B Environ* 2017;202:156–64. https://doi.org/10.1016/j.apcatb.2016.09.008.

265. Reszczyńska J, Grzyb T, Sobczak JW, Lisowski W, Gazda M, Ohtani B, et al. Lanthanide co-doped TiO_2: The effect of metal type and amount on surface properties and photocatalytic activity. *Appl Surf Sci* 2014;307:333–45. https://doi.org/10.1016/j.apsusc.2014.03.199.

266. Li M, Luan J, Zhang Y, Jiang F, Zhou X, Tang J, et al. Spectroscopic properties of Er/Yb co-doped glass ceramics containing nanocrystalline $Bi_2ZnB_2O_7$ for broadband near-infrared emission. *Ceram Int* 2019;45:18831–7. https://doi.org/10.1016/j.ceramint.2019.06.116.

267. Xin M. High luminescent TiO_2-Yb_2O_3: Er, Li complex nano spherical upconversion phosphor prepared by a hydrothermally treatment. *J Lumin* 2019;213:415–20. https://doi.org/10.1016/j.jlumin.2019.05.039.

268. Yu X, Ji Q, Zhang J, Nie Z, Yang H. Photocatalytic degradation of diesel pollutants in seawater under visible light. *Reg Stud Mar Sci* 2018;18:139–44. https://doi.org/10.1016/j.rsma.2018.02.006.

269. Yu JC, Ho W, Yu J, Yip H, Po KW, Zhao J. Efficient visible-light-induced photocatalytic disinfection on sulfur-doped nanocrystalline titania. *Environ Sci Technol* 2005;39:1175–9. https://doi.org/10.1021/es035374h.

270. Shi J wen, Zheng J tang, Wu P. Preparation, characterization and photocatalytic activities of holmium-doped titanium dioxide nanoparticles. *J Hazard Mater* 2009;161:416–22. https://doi.org/10.1016/j.jhazmat.2008.03.114.

271. Zhou W, He Y. Ho/TiO_2 nanowires heterogeneous catalyst with enhanced photocatalytic properties by hydrothermal synthesis method. *Chem Eng J* 2012;179:412–6. https://doi.org/10.1016/j.cej.2011.10.094.

272. Li JG, Wang XH, Kamiyama H, Ishigaki T, Sekiguchi T. RF plasma processing of Er-doped TiO_2 luminescent nanoparticles. *Thin Solid Films* 2006;506–507:292–6. https://doi.org/10.1016/j.tsf.2005.08.093.

273. Heshan CA, Guoguang LI, Lü W, Xiaoxia LI, Lin YU, Daguang LI. Effect of Ho-doping on photocatalytic activity of nanosized TiO_2 catalyst. *J Rare Earths* 2008;26:71–5. https://doi.org/10.1016/S1002-0721(08)60040-X.

274. Shi JW, Zheng JT, Hu Y, Zhao YC. Influence of Fe^{3+} and Ho^{3+} co-doping on the photocatalytic activity of TiO_2. *Mater Chem Phys* 2007;106:247–9. https://doi.org/10.1016/j.matchemphys.2007.05.042.

275. Navas J, Sánchez-Coronilla A, Aguilar T, De Los Santos DM, Hernández NC, Alcántara R, et al. Thermo-selective $Tm_xTi_{1-x}O_{2-x/2}$ nanoparticles: From Tm-doped anatase TiO_2 to a rutile/pyrochlore $Tm_2Ti_2O_7$ mixture. An experimental and theoretical study with a photocatalytic application. *Nanoscale* 2014;6:12740–57. https://doi.org/10.1039/c4nr03715d.

276. de los Santos DM, Navas J, Aguilar T, Sánchez-Coronilla A, Fernández-Lorenzo C, Alcántara R, et al. Tm-doped TiO_2 and $Tm_2Ti_2O_7$ pyrochlore nanoparticles: Enhancing the photocatalytic activity of rutile with a pyrochlore phase. *Beilstein J Nanotechnol* 2015;6:605–16. https://doi.org/10.3762/bjnano.6.62.

277. Fan L, Dongmei J, Xueming M. The effect of milling atmospheres on photocatalytic property of Fe-doped TiO_2 synthesized by mechanical alloying. *J Alloys Compd* 2009;470:375–8. https://doi.org/10.1016/j.jallcom.2008.02.067.

278. De Los Santos DM, Navas J, Aguilar T, Sánchez-Coronilla A, Alcántara R, Fernández-Lorenzo C, et al. Study of thulium doping effect and enhancement of photocatalytic activity of rutile TiO_2 nanoparticles. *Mater Chem Phys* 2015;161:175–84. https://doi.org/10.1016/j.matchemphys.2015.05.034.

279. Sanjay P, Chinnasamy E, Deepa K, Madhavan J, Senthil S. Synthesis, structural, morphological and optical characterization of TiO_2 and Nd^{3+} doped TiO_2 nanoparticles by sol gel method: A comparative study for photovoltaic application. *IOP Conf Ser Mater Sci Eng* 2018;360. https://doi.org/10.1088/1757-899X/360/1/012011.

280. Gomez V, Balu AM, Serrano-Ruiz JC, Irusta S, Dionysiou DD, Luque R, et al. Microwave-assisted mild-temperature preparation of neodymium-doped titania for the improved photodegradation of water contaminants. *Appl Catal A Gen* 2012;441–442:47–53. https://doi.org/10.1016/j.apcata.2012.07.003.

281. Shao X, Pan F, Zheng L, Zhang R, Zhang WY. Nd-doped TiO_2-C hybrid aerogels and their photocatalytic properties. *Xinxing Tan Cailiao/New Carbon Mater* 2018;33:116–24. https://doi.org/10.1016/S1872-5805(18)60329-4.

282. Soleimanzadeh H, Niaei A, Salari D, Mousavi SM, Tarjamannejad A. Performance study f V_2O_5/TiO_2 mixed metal oxide nanocatalysts in selective catalytic reduction of Nox prepared by co-precipitation method. *Procedia Mater Sci* 2015;11:655–60. https://doi.org/10.1016/j.mspro.2015.11.015.

283. Parnicka P, Mazierski P, Grzyb T, Lisowski W, Kowalska E, Ohtani B, et al. Influence of the preparation method on the photocatalytic activity of Nd-modified TiO_2. *Beilstein J Nanotechnol* 2018;9:447–59. https://doi.org/10.3762/bjnano.9.43.

284. Wojcieszak D, Mazur M, Kurnatowska M, Kaczmarek D, Domaradzki J, Kepinski L, et al. Influence of Nd-doping on photocatalytic properties of TiO_2 nanoparticles and thin film coatings. *Int J Photoenergy* 2014;2014. https://doi.org/10.1155/2014/463034.

285. Wang C, Ao Y, Wang P, Hou J, Qian J. Preparation, characterization and photocatalytic activity of the neodymium-doped TiO_2 hollow spheres. *Appl Surf Sci* 2010;257:227–31. https://doi.org/10.1016/j.apsusc.2010.06.071.

286. Parnicka P, Mazierski P, Grzyb T, Wei Z, Kowalska E, Ohtani B, et al. Preparation and photocatalytic activity of Nd-modified TiO_2 photocatalysts: Insight into the excitation mechanism under visible light. *J Catal* 2017;353:211–22. https://doi.org/10.1016/j.jcat.2017.07.017.

287. Rengaraj S, Venkataraj S, Yeon JW, Kim Y, Li XZ, Pang GKH. Preparation, characterization and application of Nd-TiO_2 photocatalyst for the reduction of Cr(VI) under UV light illumination. *Appl Catal B Environ* 2007;77:157–65. https://doi.org/10.1016/j.apcatb.2007.07.016.

288. Vieira GB, Scaratti G, Rodembusch FS, De Amorim SM, Peterson M, Puma GL, et al. Tuning the photoactivity of TiO_2 nanoarchitectures doped with cerium or neodymium and application to colour removal from wastewaters. *Environ Technol (United Kingdom)* 2019;0:1–15. https://doi.org/10.1080/09593330.2019.1651402.

289. Liang J, Wang J, Song K, Wang X, Yu K, Liang C. Enhanced photocatalytic activities of Nd-doped TiO_2 under visible light using a facile sol-gel method. *J Rare Earths* 2019;2–10. https://doi.org/10.1016/j.jre.2019.07.008.

290. Wu X, Yin S, Dong Q, Guo C, Kimura T, Matsushita JI, et al. Photocatalytic properties of Nd and C Codoped TiO_2 with the Whole range of visible light absorption. *J Phys Chem C* 2013;117:8345–52. https://doi.org/10.1021/jp402063n.

291. Nassoko D, Li YF, Li JL, Li X, Yu Y. Neodymium-doped TiO_2 with anatase and brookite two phases: Mechanism for photocatalytic activity enhancement under visible light and the role of electron. *Int J Photoenergy* 2012;2012. https://doi.org/10.1155/2012/716087.
292. Mazierski P, Mikolajczyk A, Bajorowicz B, Malankowska A, Zaleska-Medynska A, Nadolna J. The role of lanthanides in TiO_2-based photocatalysis: A review. *Appl Catal B Environ* 2018;233:301–17. https://doi.org/10.1016/j.apcatb.2018.04.019.
293. Wei Z-N, Jia C-L. First-principle calculations of the electronic and optical properties of Tm-doped anatase titanium dioxide. *Opt Eng* 2015;54:037107. https://doi.org/10.1117/1.oe.54.3.037107.
294. Bruyer E, Sayede A. Density functional calculations of the structural, electronic, and ferroelectric properties of high- k titanate $Re_2Ti_2O_7$ (Re=La and Nd). *J Appl Phys* 2010;108. https://doi.org/10.1063/1.3459891.
295. Li W, Wang Y, Lin H, Shah SI, Huang CP, Doren DJ, et al. Band gap tailoring of Nd^{3+}-doped TiO_2 nanoparticles. *Appl Phys Lett* 2003;83:4143–5. https://doi.org/10.1063/1.1627962.
296. Zhang YF, Lin W, Li Y, Ding KN, Li JQ. A theoretical study on the electronic structures of TiO_2: Effect of Hartree - Fock exchange. *J Phys Chem B* 2005;109:19270–7. https://doi.org/10.1021/jp0523625.
297. Ramakrishna P V., Pammi SVN, Samatha K. UV-visible upconversion studies of Nd^{3+} ions in lead tellurite glass. *Solid State Commun* 2013;155:21–4. https://doi.org/10.1016/j.ssc.2012.10.043.
298. Bünzli JCG. Lanthanide light for biology and medical diagnosis. *J Lumin* 2016;170:866–78. https://doi.org/10.1016/j.jlumin.2015.07.033.
299. Alban L, Monteiro WF, Diz FM, Miranda GM, Scheid CM, Zotti ER, et al. New quercetin-coated titanate nanotubes and their radiosensitization effect on human bladder cancer. *Mater Sci Eng C* 2020;110:110662. https://doi.org/10.1016/j.msec.2020.110662.
300. Li C, Liu C, Chen L, Ye Z, Zhang Y, Wang X, et al. Studies on the separation and in-situ sintering solidification of strontium by a highly-efficient titanate-based adsorbent. *Microporous Mesoporous Mater* 2019;288:109607. https://doi.org/10.1016/j.micromeso.2019.109607.
301. Ran J, Guo M, Zhong L, Fu H. In situ growth of $BaTiO_3$ nanotube on the surface of reduced graphene oxide: A lightweight electromagnetic absorber. *J Alloys Compd* 2019;773:423–31. https://doi.org/10.1016/j.jallcom.2018.09.142.
302. Saleem M, Butt MS, Maqbool A, Umer MA, Shahid M, Javaid F, et al. Percolation phenomena of dielectric permittivity of a microwave-sintered $BaTiO_3$–Ag nanocomposite for high energy capacitor. *J Alloys Compd* 2020;822. https://doi.org/10.1016/j.jallcom.2019.153525.
303. Demircivi P, Gulen B, Simsek EB, Berek D. Enhanced photocatalytic degradation of tetracycline using hydrothermally synthesized carbon fiber decorated $BaTiO_3$. *Mater Chem Phys* 2020;241:122236. https://doi.org/10.1016/j.matchemphys.2019.122236.
304. Zhang SW, Zhang BP, Li S, Li XY, Huang ZC. SPR enhanced photocatalytic properties of Au-dispersed amorphous $BaTiO_3$ nanocomposite thin films. *J Alloys Compd* 2016;654:112–9. https://doi.org/10.1016/j.jallcom.2015.09.053.
305. Stanciu CA, Cernea M, Secu EC, Aldica G, Ganea P, Trusca R. Lanthanum influence on the structure, dielectric properties and luminescence of $BaTiO_3$ ceramics

processed by spark plasma sintering technique. *J Alloys Compd* 2017;706:538–45. https://doi.org/10.1016/j.jallcom.2017.02.258.

306. Mohammadi P, Ghorbani-Shahna F, Bahrami A, Rafati AA, Farhadian M. Plasma-photocatalytic degradation of gaseous toluene using $SrTiO_3$/rGO as an efficient heterojunction for by-products abatement and synergistic effects. *J Photochem Photobiol A Chem* 2020;394:112460. https://doi.org/10.1016/j.jphotochem.2020.112460.

307. Bashiri R, Irfan MS, Mohamed NM, Sufian S, Ling LY, Suhaimi NA, et al. Hierarchically $SrTiO_3$@TiO_2@Fe_2O_3 nanorod heterostructures for enhanced photoelectrochemical water splitting. *Int J Hydrogen Energy* 2020. https://doi.org/10.1016/j.ijhydene.2020.02.106.

308. Patial S, Hasija V, Raizada P, Singh P, Khan Singh AAP, Asiri AM. Tunable photocatalytic activity of $SrTiO_3$ for water splitting: Strategies and future scenario. *J Environ Chem Eng* 2020;8:103791. https://doi.org/10.1016/j.jece.2020.103791.

309. Terraschke H, Wickleder C. UV, blue, green, yellow, red, and small: Newest developments on Eu^{2+} -doped nanophosphors. *Chem Rev* 2015;115:11352–78. https://doi.org/10.1021/acs.chemrev.5b00223.

310. Si D, Geng B, Wang S. One-step synthesis and morphology evolution of luminescent Eu^{2+} doped strontium aluminate nanostructures. *CrystEngComm* 2010;12:2722. https://doi.org/10.1039/b921613h.

311. Xu Y-F, Ma D-K, Guan M-L, Chen X-A, Pan Q-Q, Huang S-M. Controlled synthesis of single-crystal $SrAl_2O_4$: Eu^{2+}, Dy^{3+} nanosheets with long-lasting phosphorescence. *J Alloys Compd* 2010;502:38–42. https://doi.org/10.1016/j.jallcom.2010.04.186.

312. Dong H, Du S-R, Zheng X-Y, Lyu G-M, Sun L-D, Li L-D, et al. Lanthanide nanoparticles: From design toward bioimaging and therapy. *Chem Rev* 2015;115:10725–815. https://doi.org/10.1021/acs.chemrev.5b00091.

313. Darr JA, Zhang J, Makwana NM, Weng X. Continuous hydrothermal synthesis of inorganic nanoparticles: Applications and future directions. *Chem Rev* 2017;117:11125–238. https://doi.org/10.1021/acs.chemrev.6b00417.

314. Lee S-K, Choi G-J, Hwang U-Y, Koo K-K, Park T-J. Effect of molar ratio of KOH to Ti-isopropoxide on the formation of $BaTiO_3$ powders by hydrothermal method. *Mater Lett* 2003;57:2201–7. https://doi.org/10.1016/S0167-577X(02)01174-6.

315. Feng Q, Hirasawa M, Yanagisawa K. Synthesis of crystal-axis-oriented $BaTiO_3$ and anatase platelike particles by a hydrothermal soft chemical process. *Chem Mater* 2001;13:290–6. https://doi.org/10.1021/cm000411e.

316. Souza AE, Santos GTA, Barra BC, Macedo WD, Teixeira SR, Santos CM, et al. Photoluminescence of $SrTiO_3$: Influence of particle size and morphology. *Cryst Growth Des* 2012;12:5671–9. https://doi.org/10.1021/cg301168k.

317. Lu R, Yuan J, Shi H, Li B, Wang W, Wang D, et al. Morphology-controlled synthesis and growth mechanism of lead-free bismuth sodium titanate nanostructures via the hydrothermal route. *CrystEngComm* 2013;15:3984. https://doi.org/10.1039/c3ce40139a.

318. Cao Y, Zhu K, Wu Q, Gu Q, Qiu J. Hydrothermally synthesized barium titanate nanostructures from $K_2Ti_4O_9$ precursors: Morphology evolution and its growth mechanism. *Mater Res Bull* 2014;57:162–9. https://doi.org/10.1016/j.materresbull.2014.05.043.

319. Wang JE, Baek C, Jung YH, Kim DK. Surface-to-core structure evolution of gradient $BaTiO_3$-$Ba_{1-x}Sr_xTiO_3$ core-shell nanoparticles. *Appl Surf Sci* 2019;487:278–84. https://doi.org/10.1016/j.apsusc.2019.05.071.

320. Xu H, Wei S, Wang H, Zhu M, Yu R, Yan H. Preparation of shape controlled $SrTiO_3$ crystallites by sol–gel-hydrothermal method. *J Cryst Growth* 2006;292:159–64. https://doi.org/10.1016/j.jcrysgro.2006.04.089.

321. Gao H, Yang H, Wang S. Hydrothermal synthesis, growth mechanism, optical properties and photocatalytic activity of cubic $SrTiO_3$ particles for the degradation of cationic and anionic dyes. *Optik (Stuttg)* 2018;175:237–49. https://doi.org/10.1016/j.ijleo.2018.09.027.

322. Xiao Z, Zhang J, Jin L, Xia Y, Lei L, Wang H, et al. Enhanced up-conversion luminescence intensity in single-crystal $SrTiO_3$: Er^{3+} nanocubes by codoping with Yb^{3+} ions. *J Alloys Compd* 2017;724:139–45. https://doi.org/10.1016/j.jallcom.2017.07.024.

323. Sreedhar G, Sivanantham A, Baskaran T, Rajapandian R, Vengatesan S, Berchmans LJ, et al. A role of lithiated sarcosine TFSI on the formation of single crystalline $SrTiO_3$ nanocubes via hydrothermal method. *Mater Lett* 2014;133:127–31. https://doi.org/10.1016/j.matlet.2014.06.170.

324. Stoyanova D, Stambolova I, Blaskov V, Zaharieva K, Avramova I, Dimitrov O, et al. Mechanical milling of hydrothermally obtained $CaTiO_3$ powders—morphology and photocatalytic activity. *Nano-Structures & Nano-Objects* 2019;18:100301. https://doi.org/10.1016/j.nanoso.2019.100301.

325. Gonçalves RF, Lima ARF, Godinho MJ, Moura AP, Espinosa J, Longo E, et al. Synthesis of Pr^{3+}-doped $CaTiO_3$ using polymeric precursor and microwave-assisted hydrothermal methods: A comparative study. *Ceram Int* 2015;41:12841–8. https://doi.org/10.1016/j.ceramint.2015.06.121.

326. Karaphun A, Hunpratub S, Swatsitang E. Effect of annealing on magnetic properties of Fe-doped $SrTiO_3$ nanopowders prepared by hydrothermal method. *Microelectron Eng* 2014;126:42–8. https://doi.org/10.1016/j.mee.2014.05.001.

327. Pinatti IM, Mazzo TM, Gonçalves RF, Varela JA, Longo E, Rosa ILV. $CaTiO_3$ and $Ca_{1-3x}Sm_xTiO_3$: Photoluminescence and morphology as a result of hydrothermal microwave methodology. *Ceram Int* 2016;42:1352–60. https://doi.org/10.1016/j.ceramint.2015.09.074.

328. Mazzo TM, Moreira ML, Pinatti IM, Picon FC, Leite ER, Rosa ILV, et al. $CaTiO_3$: Eu^{3+} obtained by microwave assisted hydrothermal method: A photoluminescent approach. *Opt Mater (Amst)* 2010;32:990–7. https://doi.org/10.1016/j.optmat.2010.01.039.

329. Zhao P, Wang L, Bian L, Xu J, Chang A, Xiong X, et al. Growth mechanism, modified morphology and optical properties of coral-like $BaTiO_3$ architecture through CTAB assisted synthesis. *J Mater Sci Technol* 2015;31:223–8. https://doi.org/10.1016/j.jmst.2014.04.002.

330. Verma KC, Gupta V, Kaur J, Kotnala RK. Raman spectra, photoluminescence, magnetism and magnetoelectric coupling in pure and Fe doped $BaTiO_3$ nanostructures. *J Alloys Compd* 2013;578:5–11. https://doi.org/10.1016/j.jallcom.2013.05.025.

331. Feng L, Li Y, Wang G, Xu B. Controlled synthesis and novel photoluminescence properties of $BaTiO_3$: Eu^{3+}/Eu^{2+} nanocrystals. *Mater Res Bull* 2015;61:173–6. https://doi.org/10.1016/j.materresbull.2014.10.020.

332. Jayabal P, Sasirekha V, Mayandi J, Jeganathan K, Ramakrishnan V. A facile hydrothermal synthesis of $SrTiO_3$ for dye sensitized solar cell application. *J Alloys Compd* 2014;586:456–61. https://doi.org/10.1016/j.jallcom.2013.10.012.

333. Wang H, Zhao W, Zhang Y, Zhang S, Wang Z, Zhao D. A facile in-situ hydrothermal synthesis of $SrTiO_3/TiO_2$ microsphere composite. *Solid State Commun* 2016;236:27–31. https://doi.org/10.1016/j.ssc.2016.03.003.

334. Patil RP, More P V., Jain GH, Khanna PK, Gaikwad VB. $BaTiO_3$ nanostructures for H2S gas sensor: Influence of band-gap, size and shape on sensing mechanism. *Vacuum* 2017;146:455–61. https://doi.org/10.1016/j.vacuum.2017.08.008.

335. Petcharoen K, Sirivat A. Synthesis and characterization of magnetite nanoparticles via the chemical co-precipitation method. *Mater Sci Eng B* 2012;177:421–7. https://doi.org/10.1016/j.mseb.2012.01.003.

336. Rane AV, Kanny K, Abitha VK, Thomas S. Methods for synthesis of nanoparticles and fabrication of nanocomposites. *Synthesis of Inorganic Nanomaterials*, edited by Bhagyaraj S, Oluwafemi O, Kalarikkal N, Thomas S. Elsevier Publications; 2018, pp. 121–39. https://doi.org/10.1016/B978-0-08-101975-7.00005-1.

337. Athar T. Smart precursors for smart nanoparticles. *Emerging Nanotechnologies for Manufacturing*, Elsevier; 2015, pp. 444–538. https://doi.org/10.1016/B978-0-323-28990-0.00017-8.

338. Dembski S, Schneider C, Christ B, Retter M. Core-shell nanoparticles and their use for in vitro and in vivo diagnostics. *Core-Shell Nanostructures Drug Deliv. Theranostics*, Elsevier; 2018, p. 119–41. https://doi.org/10.1016/B978-0-08-102198-9.00005-3.

339. Nawaz M, Sliman Y, Ercan I, Lima-Tenório MK, Tenório-Neto ET, Kaewsaneha C, et al. Magnetic and pH-responsive magnetic nanocarriers. *Stimuli Responsive Polymeric Nanocarriers for Drug Delivery Applications*, Elsevier; 2019, pp. 37–85. https://doi.org/10.1016/B978-0-08-101995-5.00002-7.

340. Xu X, Liu G, Randorn C, Irvine JTS. $g-C_3N_4$ coated $SrTiO_3$ as an efficient photocatalyst for H_2 production in aqueous solution under visible light irradiation. *Int J Hydrogen Energy* 2011;36:13501–7. https://doi.org/10.1016/j.ijhydene.2011.08.052.

341. Tiwari SP, Mahata MK, Kumar K, Rai VK. Enhanced temperature sensing response of upconversion luminescence in ZnO–$CaTiO_3$: Er^{3+}/Yb^{3+} nano-composite phosphor. *Spectrochim Acta Part A Mol Biomol Spectrosc* 2015;150:623–30. https://doi.org/10.1016/j.saa.2015.05.081.

342. Zhang X, Zhang J, Ren X, Wang X-J. The dependence of persistent phosphorescence on annealing temperatures in $CaTiO_3$: Pr^{3+} nanoparticles prepared by a coprecipitation technique. *J Solid State Chem* 2008;181:393–8. https://doi.org/10.1016/j.jssc.2007.11.022.

343. Kumari A, Rai VK, Kumar K. Yellow–orange upconversion emission in Eu^{3+}–Yb^{3+} codoped $BaTiO_3$ phosphor. *Spectrochim Acta Part A Mol Biomol Spectrosc* 2014;127:98–101. https://doi.org/10.1016/j.saa.2014.02.023.

344. Mahata MK, Kumar K, Rai VK. Structural and optical properties of Er^{3+}/Yb^{3+} doped barium titanate phosphor prepared by co-precipitation method. *Spectrochim Acta Part A Mol Biomol Spectrosc* 2014;124:285–91. https://doi.org/10.1016/j.saa.2014.01.014.

345. Rao BG, Mukherjee D, Reddy BM. Novel approaches for preparation of nanoparticles. *Nanostructures for Novel Therapy: Synthesis Characterizations and Apllications*, edited by Ficai D and Grumezescu A. Elsevier Publications; 2017, pp. 1–36. https://doi.org/10.1016/B978-0-323-46142-9.00001-3.

346. Baraket L, Ghorbel A. Control preparation of aluminium chromium mixed oxides by Sol-Gel process. *Studies in Surface Science and Catalysis*, Elsevier; 1998, pp. 657–67. https://doi.org/10.1016/S0167-2991(98)80233-4.

347. Phulé PP, Wood TE. Ceramics and glasses, sol–gel synthesis of. *Encyclopedia of Materials: Science and Technology*, Elsevier; 2001, pp. 1090–5. https://doi.org/10.1016/B0-08-043152-6/00201-1.

348. Sakka Sumio. Sol–gel process and applications. *Handbook of Advanced Ceramics*, Elsevier Publications; 2013, pp. 883–910. https://doi.org/10.1016/B978-0-12-385469-8.00048-4.

349. Mitra A, De G. Sol-gel synthesis of metal nanoparticle incorporated oxide films on glass. In *Glass Nanocomposites*, edited by Karmakar B, Rademann K, Stepanov A. Elsevier; 2016, pp. 145–63. https://doi.org/10.1016/B978-0-323-39309-6.00006-7.

350. Cernea M, Monnereau O, Llewellyn P, Tortet L, Galassi C. Sol–gel synthesis and characterization of Ce doped-BaTiO$_3$. *J Eur Ceram Soc* 2006;26:3241–6. https://doi.org/10.1016/j.jeurceramsoc.2005.09.039.

351. Hao Y, Zhang J, Bi M, Feng Z, Bi K. Hollow-sphere SrTiO$_3$ nanocube assemblies with enhanced room-temperature photoluminescence. *Mater Des* 2018;155:257–63. https://doi.org/10.1016/j.matdes.2018.06.006.

352. Sun W, Gu Y, Zhang Q, Li Y, Wang H. CaTiO$_3$:Eu^{3+} layers coated SiO$_2$ particles: Core-shell structured red phosphors for near-UV white LEDs. *J Alloys Compd* 2010;493:561–4. https://doi.org/10.1016/j.jallcom.2009.12.155.

353. Gholamrezaei S, Salavati-Niasari M. An efficient dye sensitized solar cells based on SrTiO$_3$ nanoparticles prepared from a new amine-modified sol-gel route. *J Mol Liq* 2017;243:227–35. https://doi.org/10.1016/j.molliq.2017.08.031.

354. Yang Z, Zhu K, Song Z, Zhou D, Yin Z, Yan L, et al. Significant reduction of upconversion emission in CaTiO$_3$: Yb, Er inverse opals. *Thin Solid Films* 2011;519:5696–9. https://doi.org/10.1016/j.tsf.2011.03.027.

355. He X, Dong W, Zheng F, Fang L, Shen M. Effect of tartaric acid on the microstructure and photoluminescence of SrTiO$_3$: Pr^{3+} phosphors prepared by a sol–gel method. *Mater Chem Phys* 2010;123:284–8. https://doi.org/10.1016/j.matchemphys.2010.04.012.

356. Lemański K, Gągor A, Kurnatowska M, Pązik R, Dereń PJ. Spectroscopic properties of Nd^{3+} ions in nano-perovskite CaTiO$_3$. *J Solid State Chem* 2011;184:2713–8. https://doi.org/10.1016/j.jssc.2011.08.004.

357. Đorđević V, Brik MG, Srivastava AM, Medić M, Vulić P, Glais E, et al. Luminescence of Mn^{4+} ions in CaTiO$_3$ and MgTiO$_3$ perovskites: Relationship of experimental spectroscopic data and crystal field calculations. *Opt Mater (Amst)* 2017;74:46–51. https://doi.org/10.1016/j.optmat.2017.03.021.

358. Vega M, Martin IR, Llanos J. Near-infrared to visible upconversion and second harmonic generation in BaTiO$_3$: Ho^{3+} and BaTiO$_3$: Ho^{3+}/Yb^{3+} phosphors. *J Alloys Compd* 2019;806:1146–52. https://doi.org/10.1016/j.jallcom.2019.07.311.

359. Vega M, Fuentes S, Martín IR, Llanos J. Up-conversion photoluminescence of BaTiO$_3$ doped with Er^{3+} under excitation at 1500 nm. *Mater Res Bull* 2017;86:95–100. https://doi.org/10.1016/j.materresbull.2016.10.001.

360. Li J, Kuwabara M. Preparation and luminescent properties of Eu-doped BaTiO$_3$ thin films by sol–gel process. *Sci Technol Adv Mater* 2003;4:143–8. https://doi.org/10.1016/S1468-6996(03)00027-5.

361. Zhang HX, Kam CH, Zhou Y, Ng SL, Lam YL, Buddhudu S. Green up-conversion emission in Er^{3+}: BaTiO$_3$ sol-gel powders. *Spectrochim Acta Part A Mol Biomol Spectrosc* 2000;56:2231–4. https://doi.org/10.1016/S1386-1425(00)00276-6.

362. Cernea M, Vasile BS, Boni A, Iuga A. Synthesis, structural characterization and dielectric properties of Nb doped BaTiO$_3$/SiO$_2$ core–shell heterostructure. *J Alloys Compd* 2014;587:553–9. https://doi.org/10.1016/j.jallcom.2013.10.228.

363. Bang JH, Suslick KS. Applications of ultrasound to the synthesis of nano-structured materials. *Adv Mater* 2010;22:1039–59. https://doi.org/10.1002/adma.200904093.

364. Suslick KS. Sonochemistry. *Science* 1990;247:1439–45. https://doi.org/10.1126/science.247.4949.1439.

365. Flint EB, Suslick KS. The temperature of cavitation. *Science* 1991;253:1397–9. https://doi.org/10.1126/science.253.5026.1397.

366. Dutta DP, Ballal A, Nuwad J, Tyagi AK. Optical properties of sonochemically synthesized rare earth ions doped $BaTiO_3$ nanophosphors: Probable candidate for white light emission. *J Lumin* 2014;148:230–7. https://doi.org/10.1016/j.jlumin.2013.11.071.

367. Dhanalakshmi M, Nagabhushana H, Darshan GP, Basavaraj RB, Daruka Prasad B. Sonochemically assisted hollow/solid $BaTiO_3$: Dy^{3+} microspheres and their applications in effective detection of latent fingerprints and lip prints. *J Sci Adv Mater Devices* 2017;2:22–33. https://doi.org/10.1016/j.jsamd.2017.02.004.

368. Dang F, Kato K, Imai H, Wada S, Haneda H, Kuwabara M. Growth of $BaTiO_3$ nanoparticles in ethanol–water mixture solvent under an ultrasound-assisted synthesis. *Chem Eng J* 2011;170:333–7. https://doi.org/10.1016/j.cej.2011.03.076.

369. Yeshodamma S, Sunitha DV, Basavaraj RB, Darshan GP, Prasad BD, Nagabhushana H. Monovalent ions co-doped $SrTiO_3$: Pr^{3+} nanostructures for the visualization of latent fingerprints and can be red component for solid state devices. *J Lumin* 2019;208:371–87. https://doi.org/10.1016/j.jlumin.2018.12.044.

370. Dhanalakshmi M, Nagabhushana H, Darshan GP, Daruka Prasad B. Ultrasound assisted sonochemically engineered effective red luminescent labeling agent for high resolution visualization of latent fingerprints. *Mater Res Bull* 2018;98:250–64. https://doi.org/10.1016/j.materresbull.2017.09.059.

371. Yashodamma S, Darshan GP, Basavaraj RB, Udayabhanu, Nagabhushana H. Ultrasound assisted fabrication of $SrTiO_3$ nanopowders: Effect of electron beam induced structural and luminescence properties for solid state lightning and high temperature dosimetry applications. *Opt Mater (Amst)* 2019;92:386–98. https://doi.org/10.1016/j.optmat.2019.04.030.

372. Sandhyarani A, Basavaraj RB, Darshan GP, Nagabhushana H, Kokila MK. Influence of surface modification on enhancement of luminescent properties of SiO_2@$SrTiO_3$: Dy^{3+} nanopowders: Probe for visualization of sweat pores present in latent fingerprints. *Optik (Stuttg)* 2019;181:1139–57. https://doi.org/10.1016/j.ijleo.2018.12.088.

373. Zeng F, Cao M, Zhang L, Liu M, Hao H, Yao Z, et al. Microstructure and dielectric properties of $SrTiO_3$ ceramics by controlled growth of silica shells on $SrTiO_3$ nanoparticles. *Ceram Int* 2017;43:7710–6. https://doi.org/10.1016/j.ceramint.2017.03.073.

374. Utara S, Hunpratub S. Ultrasonic assisted synthesis of $BaTiO_3$ nanoparticles at 25°C and atmospheric pressure. *Ultrason Sonochem* 2018;41:441–8. https://doi.org/10.1016/j.ultsonch.2017.10.008.

375. Kingsley JJ, Suresh K, Patil KC. Combustion synthesis of fine-particle metal aluminates. *J Mater Sci* 1990;25:1305–12. https://doi.org/10.1007/BF00585441.

376. Varma A, Mukasyan AS, Rogachev AS, Manukyan K V. Solution combustion synthesis of nanoscale materials. *Chem Rev* 2016;116:14493–586. https://doi.org/10.1021/acs.chemrev.6b00279.

377. Mukasyan AS, Dinka P. Novel approaches to solution-combustion synthesis of nanomaterials. *Int J Self-Propagating High-Temperature Synth* 2007;16:23–35. https://doi.org/10.3103/S1061386207010049.

378. Yeh C-L. Combustion synthesis: Principles and applications. *Reference Module in Materials Science and Materials Engineering*, Elsevier Publications; 2016. https://doi.org/10.1016/B978-0-12-803581-8.03743-7.

379. Yin Q, Qiu K, Zhang W, Chen X, Zhang P, Tang Q, et al. Crystal structure and luminescence properties of $CaTiO_3$: Dy^{3+} phosphor co-doped with Zr^{4+}. *Opt Mater (Amst)* 2019;98:109446. https://doi.org/10.1016/j.optmat.2019.109446.

380. Bhagya NP, Prashanth PA, Krishna RH, Nagabhushana BM, Raveendra RS. Photoluminescence studies of Eu^{3+} activated $SrTiO_3$ nanophosphor prepared by solution combustion approach. *Optik (Stuttg)* 2017;145:678–87. https://doi.org/10.1016/j.ijleo.2017.07.003.

381. Sandhyarani A, Kokila MK, Darshan GP, Basavaraj RB, Daruka Prasad B, Sharma SC, et al. Versatile core–shell SiO_2 @$SrTiO_3$: Eu^{3+}, Li^+ nanopowders as fluorescent label for the visualization of latent fingerprints and anti-counterfeiting applications. *Chem Eng J* 2017;327:1135–50. https://doi.org/10.1016/j.cej.2017.06.093.

382. Shivram M, Prashantha SC, Nagabhushana H, Sharma SC, Thyagarajan K, Harikrishna R, et al. $CaTiO_3$: Eu^{3+} red nanophosphor: Low temperature synthesis and photoluminescence properties. *Spectrochim Acta Part A Mol Biomol Spectrosc* 2014;120:395–400. https://doi.org/10.1016/j.saa.2013.09.114.

383. Inamdar HK, Ambika Prasad MVN, Basavaraj RB, Sasikala M, Sharma SC, Nagabhushana H. Promising red emission from functionalized Polypyrrole/$CaTiO_3$: Eu^{3+} nano-composites for photonic applications. *Opt Mater* 2019;88:458–65. https://doi.org/10.1016/j.optmat.2018.11.042.

384. Marí B, Singh KC, Cembrero-Coca P, Singh I, Singh D, Chand S. Red emitting $MTiO_3$ (M=Ca or Sr) phosphors doped with Eu^{3+} or Pr^{3+} with some cations as co-dopants. *Displays* 2013;34:346–51. https://doi.org/10.1016/j.displa.2013.07.003.

385. Dhanalakshmi M, Basavaraj RB, Darshan GP, Sharma SC, Nagabhushana H. Pivotal role of fluxes in $BaTiO_3$: Eu^{3+} nano probes for visualization of latent fingerprints on multifaceted substrates and anti-counterfeiting applications. *Microchem J* 2019;145:226–34. https://doi.org/10.1016/j.microc.2018.10.020.

386. Bhagya NP, Prashanth PA, Raveendra RS, Sathyanarayani S, Ananda S, Nagabhushana BM, et al. Adsorption of hazardous cationic dye onto the combustion derived $SrTiO_3$ nanoparticles: Kinetic and isotherm studies. *J Asian Ceram Soc* 2016;4:68–74. https://doi.org/10.1016/j.jascer.2015.11.005.

387. Shivaram M, Krishna RH, Nagabhushana H, Sharma SC, Nagabhushana BM, Ravikumar BS, et al. Synthesis, characterization, EPR and thermoluminescence properties of $CaTiO_3$ nanophosphor. *Mater Res Bull* 2013;48:1490–8. https://doi.org/10.1016/j.materresbull.2012.12.065.

388. Shivaram M, Nagabhushana H, Sharma SC, Prashantha SC, Daruka Prasad B, Dhananjaya N, et al. Synthesis and luminescence properties of Sm^{3+} doped $CaTiO_3$ nanophosphor for application in white LED under NUV excitation. *Spectrochim Acta Part A Mol Biomol Spectrosc* 2014;128:891–901. https://doi.org/10.1016/j.saa.2014.02.117.

389. Zhao Jingping, Yan Wenfu. Microwave-assisted inorganic syntheses. *Modern Inorganic Synthetic Chemistry*, Elsevier Publications; 2011, pp. 173–95. https://doi.org/10.1016/B978-0-444-53599-3.10008-3.

390. Xu F, Yan H, Chen J, Zhang Z, Fan C. Nanoscale Co_3O_4 powders prepared by an enhanced solid-state reaction method. *Ceram Int* 2020. https://doi.org/10.1016/j.ceramint.2020.02.183.
391. Huang X, Chen J, Xu F. Structural and magnetic properties of Fe-doped SnS_2 nanopowders synthesized by solid-state reaction method. *Chem Phys Lett* 2020;739:137042. https://doi.org/10.1016/j.cplett.2019.137042.
392. Li F, Liu X, He T. Solid state synthesis of $CaTiO_3$: Dy^{3+}/Eu^{3+} phosphors towards white light emission. *Chem Phys Lett* 2017;686:78–82. https://doi.org/10.1016/j.cplett.2017.08.042.
393. Liu P, Yin J, Mi X, Zhang L, Bie L. Enhanced photoluminescence of $CaTiO_3$: Eu^{3+} red phosphors prepared by H_3BO_3 assisted solid state synthesis. *J Rare Earths* 2013;31:555–8. https://doi.org/10.1016/S1002-0721(12)60319-6.
394. Tian X, Wang C, Wen J, Lian S, Ji C, Huang Z, et al. High temperature sensitivity phosphor based on an old material: Red emitting H_3BO_3 flux assisted $CaTiO_3$: Pr^{3+}. *J Lumin* 2019;214:116528. https://doi.org/10.1016/j.jlumin.2019.116528.
395. Tian X, Lian S, Ji C, Huang Z, Wen J, Chen Z, et al. Enhanced photoluminescence and ultrahigh temperature sensitivity from NaF flux assisted $CaTiO_3$: Pr^{3+} red emitting phosphor. *J Alloys Compd* 2019;784:628–40. https://doi.org/10.1016/j.jallcom.2019.01.087.
396. Sukul PP, Mahata MK, Ghorai UK, Kumar K. Crystal phase induced upconversion enhancement in Er^{3+}/Yb^{3+} doped $SrTiO_3$ ceramic and its temperature sensing studies. *Spectrochim Acta Part A Mol Biomol Spectrosc* 2019;212:78–87. https://doi.org/10.1016/j.saa.2018.12.039.
397. Zhang J, Huang M, Yanagisawa K, Yao S. $NaCl–H_2O$-assisted preparation of $SrTiO_3$ nanoparticles by solid state reaction at low temperature. *Ceram Int* 2015;41:5439–44. https://doi.org/10.1016/j.ceramint.2014.12.110.
398. Ha MG, Byeon MR, Hong TE, Bae JS, Kim Y, Park S, et al. Sm^{3+}-doped $CaTiO_3$ phosphor: Synthesis, structure, and photoluminescent properties. *Ceram Int* 2012;38:1365–70. https://doi.org/10.1016/j.ceramint.2011.09.015.
399. Jyothi G, Gopchandran KG. Compositional tuning and site selective excitations in $SrTiO_3$: Y^{3+}, Eu^{3+} red phosphors. *Dye Pigment* 2018;149:531–42. https://doi.org/10.1016/j.dyepig.2017.10.040.
400. Chen R, Song F, Chen D, Peng Y. Improvement of the luminescence properties of $CaTiO_3$: Pr obtained by modified solid-state reaction. *Powder Technol* 2009;194:252–5. https://doi.org/10.1016/j.powtec.2009.05.005.
401. Suriyamurthy N, Panigrahi BS. Investigations on luminescence of rare earths doped $CaTiO_3$: Pr^{3+} phosphor. *J Rare Earths* 2010;28:488–92. https://doi.org/10.1016/S1002-0721(09)60138-1.
402. Qin X, Li Y, Li Y, Wu Y, Chen R, Sharafudeen K, et al. Inducing NIR long persistent phosphorescence in Cr-doped perovskite titanate via redox. *J Alloys Compd* 2016;666:387–91. https://doi.org/10.1016/j.jallcom.2016.01.135.
403. Orihashi T, Adachi S. Synthesis condition and structural/luminescent properties of $CaTiO_3$: Eu^{3+} red phosphor. *J Alloys Compd* 2015;646:1116–23. https://doi.org/10.1016/j.jallcom.2015.06.200.
404. Selvarajan S, Alluri NR, Chandrasekhar A, Kim S-J. Direct detection of cysteine using functionalized $BaTiO_3$ nanoparticles film based self-powered biosensor. *Biosens Bioelectron* 2017;91:203–10. https://doi.org/10.1016/j.bios.2016.12.006.
405. Zhang L, Pan H, Liu H, Zhang B, Jin L, Zhu M, et al. Theoretical spectra identification and fluorescent properties of reddish orange Sm-doped $BaTiO_3$ phosphors. *J Alloys Compd* 2015;643:247–52. https://doi.org/10.1016/j.jallcom.2015.04.087.

406. Selvarajan S, Alluri NR, Chandrasekhar A, Kim S-J. Unconventional active biosensor made of piezoelectric $BaTiO_3$ nanoparticles for biomolecule detection. *Sensors Actuators B Chem* 2017;253:1180–7. https://doi.org/10.1016/j.snb.2017.07.159.

407. Kumar Singh D, Mondal K, Manam J. Improved photoluminescence, thermal stability and temperature sensing performances of K^+ incorporated perovskite $BaTiO_3$: Eu^{3+} red emitting phosphors. *Ceram Int* 2017;43:13602–11. https://doi.org/10.1016/j.ceramint.2017.07.069.

408. Singh DK, Manam J. Efficient dual emission mode of green emitting perovskite $BaTiO_3$: Er^{3+} phosphors for display and temperature sensing applications. *Ceram Int* 2018;44:10912–20. https://doi.org/10.1016/j.ceramint.2018.03.151.

409. Fu J, Zhang Q, Li Y, Wang H. Preparation and photoluminescence characteristics of a new promising red NUV phosphor $CaTiO_3$: Eu^{3+}. *J Alloys Compd* 2009;485:418–21. https://doi.org/10.1016/j.jallcom.2009.05.128.

410. Yin S, Chen D, Tang W, Peng Y. Synthesis of $CaTiO_3$: Pr persistent phosphors by a modified solid-state reaction. *Mater Sci Eng B* 2007;136:193–6. https://doi.org/10.1016/j.mseb.2006.09.014.

411. Noto LL, Pitale SS, Gusowki MA, Terblans JJ, Ntwaeaborwa OM, Swart HC. Afterglow enhancement with In^{3+} codoping in $CaTiO_3$: Pr^{3+} red phosphor. *Powder Technol* 2013;237:141–6. https://doi.org/10.1016/j.powtec.2013.01.029.

412. Ryu H, Singh BK, Bartwal KS, Brik MG, Kityk IV. Novel efficient phosphors on the base of Mg and Zn co-doped $SrTiO_3$: Pr^{3+}. *Acta Mater* 2008;56:358–63. https://doi.org/10.1016/j.actamat.2007.09.041.

413. Best SM, Marti PC. Mineral coatings for orthopaedic applications. *Coatings for Biomedical Applications*, Elsevier Publications; 2012, pp. 43–74. https://doi.org/10.1533/9780857093677.1.43.

414. Siva Prasanna S.R.V., Balaji K., Pandey Shyam, Rana Shravendra. Metal oxide based nanomaterials and their polymer nanocomposites. *Nanomaterials and Polymer Nanocomposites*, Elsevier; 2019, pp. 123–44. https://doi.org/10.1016/B978-0-12-814615-6.00004-7.

415. Michelmore A. Thin film growth on biomaterial surfaces. *Thin Film Coatings for Biomaterials and Biomedical Applications*, Elsevier; 2016, pp. 29–47. https://doi.org/10.1016/B978-1-78242-453-6.00002-X.

416. Yang HK, Chung JW, Moon BK, Choi BC, Jeong JH, Jang K, et al. Photoluminescence characteristics of Li-doped $CaTiO_3$: Pr^{3+} thin films grown on Si (100) substrate by PLD. *Thin Solid Films* 2010;518:6219–22. https://doi.org/10.1016/j.tsf.2010.03.171.

417. Xiong ZW, Cao LH. Red-ultraviolet photoluminescence tuning by Ni nanocrystals in epitaxial $SrTiO_3$ matrix. *Appl Surf Sci* 2018;445:65–70. https://doi.org/10.1016/j.apsusc.2018.03.179.

418. Tsuchiya T, Nakajima T, Shinoda K. Improvement of the photoluminescence of $CaTiO_3$: Pr phosphor film grown by excimer laser-assisted metal organic decomposition. *Mater Lett* 2013;108:176–8. https://doi.org/10.1016/j.matlet.2013.06.092.

419. Fasasi AY, Ngom BD, Kana-Kana JB, Bucher R, Maaza M, Theron C, et al. Synthesis and characterisation of Gd-doped $BaTiO_3$ thin films prepared by laser ablation for optoelectronic applications. *J Phys Chem Solids* 2009;70:1322–9. https://doi.org/10.1016/j.jpcs.2009.06.022.

420. Zafar MS, Farooq I, Awais M, Najeeb S, Khurshid Z, Zohaib S. Bioactive surface coatings for enhancing osseointegration of dental implants. In *Biomedical, Therapeutic and Clinical Applications of Bioactive Glasses*, edited by Kaur G. Elsevier; 2019, pp. 313–29. https://doi.org/10.1016/B978-0-08-102196-5.00011-2.

421. Depla D, Mahieu S, Greene JE. Sputter deposition processes. *Handbook of Deposition Technologies for Films and Coatings*, Elsevier; 2010, pp. 253–296. https://doi.org/10.1016/B978-0-8155-2031-3.00005-3.

422. Mattox DM. Introduction. *Handbook of Physical Vapor Deposition (PVD) Processing*, Elsevier; 2010, pp. 1–24. https://doi.org/10.1016/B978-0-8155-2037-5.00001-0.

423. Dawber M. Sputtering techniques for epitaxial growth of complex oxides. *Epitaxial Growth of Complex Metal Oxides*, Elsevier; 2015, pp. 31–45. https://doi.org/10.1016/B978-1-78242-245-7.00002-6.

424. Sarakha L, Bégou T, Goullet A, Cellier J, Bousquet A, Tomasella E, et al. Influence of synthesis conditions on optical and electrical properties of $CaTiO_3$: Pr^{3+} thin films deposited by radiofrequency sputtering for electroluminescent device. *Surf Coatings Technol* 2011;205:S250–3. https://doi.org/10.1016/j.surfcoat.2011.01.015.

425. Chung SM, Han SH, Song KH, Kim ES, Kim YJ. Luminescent properties of $CaTiO_3$: Pr thin-film phosphor deposited on ZnO/ITO/glass substrate. *J Lumin* 2005;114:227–33. https://doi.org/10.1016/j.jlumin.2005.01.007.

426. Shihab NK, Acharyya JN, Rasi UPM, Gangineni RB, Vijaya Prakash G, Narayana Rao D. Cavity enhancement in nonlinear absorption and photoluminescence of $BaTiO_3$. *Optik (Stuttg)* 2019:163896. https://doi.org/10.1016/j.ijleo.2019.163896.

427. Maneeshya LV, Thomas PV, Joy K. Effects of site substitutions and concentration on the structural, optical and visible photoluminescence properties of Er doped $BaTiO_3$ thin films prepared by RF magnetron sputtering. *Opt Mater (Amst)* 2015;46:304–9. https://doi.org/10.1016/j.optmat.2015.04.036.

428. Chen A-N, Wu J-M, Cheng L-J, Liu S-J, Ma Y-X, Li H, et al. Enhanced densification and dielectric properties of $CaTiO_{3-0.3}$ $NdAlO_3$ ceramics fabricated by direct coagulation casting. *J Eur Ceram Soc* 2019. https://doi.org/10.1016/j.jeurceramsoc.2019.12.033.

429. Paiva DVM, Silva MAS, de Oliveira RGM, Rodrigues AR, Fechine LMUD, Sombra ASB, et al. Magneto-dielectric composite based on $Y_3Fe_5O_{12}$ – $CaTiO_3$ for radio frequency and microwave applications. *J Alloys Compd* 2019;783:652–61. https://doi.org/10.1016/j.jallcom.2018.12.366.

430. Gillani SSA, Ahmad R, Islah-u-din, Rizwan M, Shakil M, Rafique M, et al. First-principles investigation of structural, electronic, optical and thermal properties of Zinc doped $SrTiO_3$. *Optik (Stuttg)* 2020;201. https://doi.org/10.1016/j.ijleo.2019.163481.

431. Wang H, Liu B, Wang X. Effects of dielectric thickness on energy storage properties of surface modified $BaTiO_3$ multilayer ceramic capacitors. *J Alloys Compd* 2019:152804. https://doi.org/10.1016/j.jallcom.2019.152804.

432. Singh DK, Manam J. Efficient dual emission mode of green emitting perovskite $BaTiO_3$: Er^{3+} phosphors for display and temperature sensing applications. *Ceram Int* 2018;44:10912–20. https://doi.org/10.1016/j.ceramint.2018.03.151.

433. Kumar Singh D, Mondal K, Manam J. Improved photoluminescence, thermal stability and temperature sensing performances of K^+ incorporated perovskite $BaTiO_3$: Eu^{3+} red emitting phosphors. *Ceram Int* 2017;43:13602–11. https://doi.org/10.1016/j.ceramint.2017.07.069.

434. Boudali A, Khodja MD, Amrani B, Bourbie D, Amara K, Abada A. First-principles study of structural, elastic, electronic, and thermal properties of $SrTiO_3$ perovskite cubic. *Phys Lett Sect A Gen At Solid State Phys* 2009;373:879–84. https://doi.org/10.1016/j.physleta.2008.12.017.

435. Garba ZN, Zhou W, Zhang M, Yuan Z. A review on the preparation, characterization and potential application of perovskites as adsorbents for wastewater treatment. *Chemosphere* 2020;244. https://doi.org/10.1016/j.chemosphere.2019.125474.

436. Bakhshi H, Sarraf-Mamoory R, Yourdkhani A, AbdelNabi AA, Mozharivskyj Y. Sol-gel synthesis, spark plasma sintering, structural characterization, and thermal conductivity measurement of heavily Nb-doped $SrTiO_3/TiO_2$ nanocomposites. *Ceram Int* 2020;46:3224–35. https://doi.org/10.1016/j.ceramint.2019.10.027.

437. Park B-G. Photoluminescence of Eu^{3+}-doped $CaTiO_3$ perovskites and their photocatalytic properties with a metal ion loading. *Chem Phys Lett* 2019;722:44–9. https://doi.org/10.1016/j.cplett.2019.03.007.

438. Ravez J, Simon A. Classical or relaxor ferroelectric ceramics in the $BaTiO_3$-$KNbO_3$-$CaTiO_3$ system. *Solid State Sci* 1999;1:25–35. https://doi.org/10.1016/S1293-2558(00)80062-X.

439. Cao F, Tian W, Wang M, Li L. Polarized ferroelectric field-enhanced self-powered perovskite photodetector. *ACS Photonics* 2018;5:3731–8. https://doi.org/10.1021/acsphotonics.8b00770.

440. Zhu XN, Gao TT, Xu X, Liang WZ, Lin Y, Chen C, et al. Piezoelectric and dielectric properties of multilayered $BaTiO_3$ /(Ba,Ca)TiO_3 /$CaTiO_3$ thin films. *ACS Appl Mater Interfaces* 2016;8:22309–15. https://doi.org/10.1021/acsami.6b05469.

441. Xiong C, Pernice WHP, Ngai JH, Reiner JW, Kumah D, Walker FJ, et al. Active silicon integrated nanophotonics: Ferroelectric $BaTiO_3$ devices. *Nano Lett* 2014;14:1419–25. https://doi.org/10.1021/nl404513p.

442. Lin M-F, Thakur VK, Tan EJ, Lee PS. Surface functionalization of $BaTiO_3$ nanoparticles and improved electrical properties of $BaTiO_3$/polyvinylidene fluoride composite. *RSC Adv* 2011;1:576. https://doi.org/10.1039/c1ra00210d.

443. Koka A, Zhou Z, Sodano HA. Vertically aligned $BaTiO_3$ nanowire arrays for energy harvesting. *Energy Environ Sci* 2014;7:288–96. https://doi.org/10.1039/C3EE42540A.

444. Manika GC, Psarras GC. $SrTiO_3$/epoxy nanodielectrics as bulk energy storage and harvesting systems: The role of conductivity. *ACS Appl Energy Mater* 2020:acsaem.9b01953. https://doi.org/10.1021/acsaem.9b01953.

445. Pei H, Zhang Y, Guo S, Ren L, Yan H, Luo B, et al. Orientation-dependent optical magnetoelectric effect in patterned $BaTiO_3$/$La_{0.67}Sr_{0.33}MnO_3$ heterostructures. *ACS Appl Mater Interfaces* 2018;10:30895–900. https://doi.org/10.1021/acsami.8b10566.

446. Erdem D, Bingham NS, Heiligtag FJ, Pilet N, Warnicke P, Vaz CAF, et al. Nanoparticle-based magnetoelectric $BaTiO_3$ –$CoFe_2O_4$ thin film heterostructures for voltage control of magnetism. *ACS Nano* 2016;10:9840–51. https://doi.org/10.1021/acsnano.6b05469.

447. Assirey EAR. Perovskite synthesis, properties and their related biochemical and industrial application. *Saudi Pharm J* 2019;27:817–29. https://doi.org/10.1016/j.jsps.2019.05.003.

448. Kamalasanan MN, Chandra S, Joshi PC, Mansingh A. Structural and optical properties of sol-gel-processed $BaTiO_3$ ferroelectric thin films. *Appl Phys Lett* 1991;59:3547–9. https://doi.org/10.1063/1.105653.

449. Kingon AI, Streiffer SK, Basceri C, Summerfelt SR. High-permittivity perovskite thin films for dynamic random-access memories. *MRS Bull* 1996;21:46–52. https://doi.org/10.1557/S0883769400035910.

450. Nenasheva EA, Kanareykin AD, Kartenko NF, Dedyk AI, Karmanenko SF. Ceramics materials based on (Ba, Sr)TiO$_3$ solid solutions for tunable microwave devices. *J Electroceramics* 2004;13:235–8. https://doi.org/10.1007/s10832-004-5104-0.

451. Dimos D, Mueller CH. Perovskite thin films for high-frequency capacitor applications. *Annu Rev Mater Sci* 1998;28:397–419. https://doi.org/10.1146/annurev.matsci.28.1.397.

452. Protesescu L, Yakunin S, Bodnarchuk MI, Krieg F, Caputo R, Hendon CH, et al. Nanocrystals of cesium lead halide perovskites (CsPbX$_3$, X=Cl, Br, and I): Novel optoelectronic materials showing bright emission with wide color gamut. *Nano Lett* 2015;15:3692–6. https://doi.org/10.1021/nl5048779.

453. Uchino K, Nomura S, Cross LE, Newnham RE, Jang SJ. Electrostrictive effect in perovskites and its transducer applications. *J Mater Sci* 1981;16:569–78. https://doi.org/10.1007/BF00552193.

454. Muralt P, Polcawich RG, Trolier-McKinstry S. Piezoelectric thin films for sensors, actuators, and energy harvesting. *MRS Bull* 2009;34:658–64. https://doi.org/10.1557/mrs2009.177.

455. Schaak RE, Mallouk TE. Prying apart Ruddlesden–Popper phases: Exfoliation into sheets and nanotubes for assembly of perovskite thin films. *Chem Mater* 2000;12:3427–34. https://doi.org/10.1021/cm000495r.

456. Ishihara T, editor. *Perovskite Oxide for Solid Oxide Fuel Cells*. Boston, MA: Springer US; 2009. https://doi.org/10.1007/978-0-387-77708-5.

457. Zheludev IS. *Physics of Crystalline Dielectrics*. Boston, MA: Springer US; 1995. https://doi.org/10.1007/978-1-4684-8076-4.

458. Samara GA. Pressure and temperature dependence of the dielectric properties and phase transitions of the ferroelectric perovskites: PbTiO$_3$ and BaTiO$_3$. *Ferroelectrics* 1971;2:277–89. https://doi.org/10.1080/00150197108234102.

459. Al-Shakarchi EK. Dielectric properties of BaTiO$_3$ ceramic prepared by freeze drying method. *J Korean Phys Soc* 2010;57:245–50. https://doi.org/10.3938/jkps.57.245.

460. Nanni P, Leoni M, Buscaglia V, Aliprandi G. Low-temperature aqueous preparation of barium metatitanate powders. *J Eur Ceram Soc* 1994;14:85–90. https://doi.org/10.1016/0955-2219(94)90048-5.

461. Masó N, Beltrán H, Cordoncillo E, Flores AA, Escribano P, Sinclair DC, et al. Synthesis and electrical properties of Nb-doped BaTiO$_3$. *J Mater Chem* 2006;16:3114–9. https://doi.org/10.1039/b601251e.

462. Luo B, Wang X, Tian E, Li G, Li L. Electronic structure, optical and dielectric properties of BaTiO$_3$ /CaTiO$_3$ /SrTiO$_3$ ferroelectric superlattices from first-principles calculations. *J Mater Chem C* 2015;3:8625–33. https://doi.org/10.1039/C5TC01622C.

463. Kim P, Doss NM, Tillotson JP, Hotchkiss PJ, Pan M-J, Marder SR, et al. High energy density nanocomposites based on surface-modified BaTiO$_3$ and a ferroelectric polymer. *ACS Nano* 2009;3:2581–92. https://doi.org/10.1021/nn9006412.

464. Tang H, Lin Y, Sodano HA. Synthesis of high aspect ratio BaTiO$_3$ nanowires for high energy density nanocomposite capacitors. *Adv Energy Mater* 2013;3:451–6. https://doi.org/10.1002/aenm.201200808.

465. Nayak S, Sahoo B, Chaki TK, Khastgir D. Facile preparation of uniform barium titanate (BaTiO$_3$) multipods with high permittivity: Impedance and temperature dependent dielectric behavior. *RSC Adv* 2014;4:1212–24. https://doi.org/10.1039/c3ra44815k.

466. Ashiri R. On the solid-state formation of BaTiO₃ nanocrystals from mechanically activated BaCO₃ and TiO₂ powders: Innovative mechanochemical processing, the mechanism involved, and phase and nanostructure evolutions. *RSC Adv* 2016;6:17138–50. https://doi.org/10.1039/C5RA22942A.

467. Huang L, Chen Z, Wilson JD, Banerjee S, Robinson RD, Herman IP, et al. Barium titanate nanocrystals and nanocrystal thin films: Synthesis, ferroelectricity, and dielectric properties. *J Appl Phys* 2006;100:034316. https://doi.org/10.1063/1.2218765.

468. Maxim F, Ferreira P, Vilarinho PM, Reaney I. Hydrothermal synthesis and crystal growth studies of BaTiO₃ using Ti nanotube precursors. *Cryst Growth Des* 2008;8:3309–15. https://doi.org/10.1021/cg800215r.

469. Dhak P, Dhak D, Das M, Pramanik K, Pramanik P. Impedance spectroscopy study of LaMnO₃ modified BaTiO₃ ceramics. *Mater Sci Eng B* 2009;164:165–71. https://doi.org/10.1016/j.mseb.2009.09.011.

470. Tang Z, Zhou Z, Zhang Z. Experimental study on the mechanism of BaTiO₃-based PTC–CO gas sensor. *Sensors Actuators B Chem* 2003;93:391–5. https://doi.org/10.1016/S0925-4005(03)00197-7.

471. Dang Z-M, Yuan J-K, Yao S-H, Liao R-J. Flexible nanodielectric materials with high permittivity for power energy storage. *Adv Mater* 2013;25:6334–65. https://doi.org/10.1002/adma.201301752.

472. Xie L, Huang X, Li B-W, Zhi C, Tanaka T, Jiang P. Core–satellite Ag@BaTiO₃ nanoassemblies for fabrication of polymer nanocomposites with high discharged energy density, high breakdown strength and low dielectric loss. *Phys Chem Chem Phys* 2013;15:17560. https://doi.org/10.1039/c3cp52799a.

473. Rao Y, Ogitani S, Kohl P, Wong CP. Novel polymer-ceramic nanocomposite based on high dielectric constant epoxy formula for embedded capacitor application. *J Appl Polym Sci* 2002;83:1084–90. https://doi.org/10.1002/app.10082.

474. Rabuffi M, Picci G. Status quo and future prospects for metallized polypropylene energy storage capacitors. *IEEE Trans Plasma Sci* 2002;30:1939–42. https://doi.org/10.1109/TPS.2002.805318.

475. Kim P, Jones SC, Hotchkiss PJ, Haddock JN, Kippelen B, Marder SR, et al. Phosphonic acid-modified barium titanate polymer nanocomposites with high permittivity and dielectric strength. *Adv Mater* 2007;19:1001–5. https://doi.org/10.1002/adma.200602422.

476. Ávila HA, Ramajo LA, Góes MS, Reboredo MM, Castro MS, Parra R. Dielectric behavior of epoxy/BaTiO₃ composites using nanostructured ceramic fibers obtained by electrospinning. *ACS Appl Mater Interfaces* 2013;5:505–10. https://doi.org/10.1021/am302646z.

477. Lin M-F, Lee PS. Formation of PVDF-g-HEMA/BaTiO₃ nanocomposites via in situ nanoparticle synthesis for high performance capacitor applications. *J Mater Chem A* 2013;1:14455. https://doi.org/10.1039/c3ta13190d.

478. Guo N, DiBenedetto SA, Tewari P, Lanagan MT, Ratner MA, Marks TJ. Nanoparticle, size, shape, and interfacial effects on leakage current density, permittivity, and breakdown strength of metal oxide–polyolefin nanocomposites: Experiment and theory. *Chem Mater* 2010;22:1567–78. https://doi.org/10.1021/cm902852h.

479. Xie L, Huang X, Yang K, Li S, Jiang P. "Grafting to" route to PVDF-HFP-GMA/BaTiO₃ nanocomposites with high dielectric constant and high thermal conductivity for energy storage and thermal management applications. *J Mater Chem A* 2014;2:5244–51. https://doi.org/10.1039/c3ta15156e.

480. Hester JF, Banerjee P, Won Y-Y, Akthakul A, Acar MH, Mayes AM. ATRP of amphiphilic graft copolymers based on PVDF and their use as membrane additives. *Macromolecules* 2002;35:7652–61. https://doi.org/10.1021/ma0122270.

481. Thakur VK, Tan EJ, Lin M-F, Lee PS. Poly(vinylidene fluoride)-graft-poly(2-hydroxyethyl methacrylate): A novel material for high energy density capacitors. *J Mater Chem* 2011;21:3751. https://doi.org/10.1039/c0jm02408b.

482. Xie L, Huang X, Huang Y, Yang K, Jiang P. Core-shell structured hyperbranched aromatic polyamide/BaTiO$_3$ hybrid filler for poly(vinylidene fluoride-trifluoroethylene-chlorofluoroethylene) nanocomposites with the dielectric constant comparable to that of percolative composites. *ACS Appl Mater Interfaces* 2013;5:1747–56. https://doi.org/10.1021/am302959n.

483. Lee H-W, Chu MSH, Lu H-Y. Phase mixture and reliability of BaTiO$_3$-based X7R multilayer ceramic capacitors: X-Ray diffractometry and Raman spectroscopy. *J Am Ceram Soc* 2011;94:1556–62. https://doi.org/10.1111/j.1551-2916.2010.04248.x.

484. Zhou T, Zha J-W, Cui R-Y, Fan B-H, Yuan J-K, Dang Z-M. Improving dielectric properties of BaTiO$_3$/ferroelectric polymer composites by employing surface hydroxylated BaTiO$_3$ nanoparticles. *ACS Appl Mater Interfaces* 2011;3:2184–8. https://doi.org/10.1021/am200492q.

485. Fu J, Hou Y, Zheng M, Wei Q, Zhu M, Yan H. Improving dielectric properties of PVDF composites by employing surface modified strong polarized BaTiO$_3$ particles derived by molten salt method. *ACS Appl Mater Interfaces* 2015;7:24480–91. https://doi.org/10.1021/acsami.5b05344.

486. Mao Y, Park T-J, Zhang F, Zhou H, Wong SS. Environmentally friendly methodologies of nanostructure synthesis. *Small* 2007;3:1122–39. https://doi.org/10.1002/smll.200700048.

487. Xing X, Zhang C, Qiao L, Liu G, Meng J. Facile preparation of ZnTiO$_3$ ceramic powders in sodium/potassium chloride melts. *J Am Ceram Soc* 2006;89:1150–2. https://doi.org/10.1111/j.1551-2916.2005.00853.x.

488. Adam J, Lehnert T, Klein G, McMeeking RM. Ferroelectric properties of composites containing BaTiO$_3$ nanoparticles of various sizes. *Nanotechnology* 2014;25:065704. https://doi.org/10.1088/0957-4484/25/6/065704.

489. Chen S, Yao K, Tay FEH, Liow CL. Ferroelectric poly(vinylidene fluoride) thin films on Si substrate with the β phase promoted by hydrated magnesium nitrate. *J Appl Phys* 2007;102:104108. https://doi.org/10.1063/1.2812702.

490. Chanmal C V., Jog JP. Dielectric relaxations in PVDF/BaTiO$_3$ nanocomposites. *Express Polym Lett* 2008;2:294–301. https://doi.org/10.3144/expresspolymlett.2008.35.

491. Wu W, Huang X, Li S, Jiang P, Toshikatsu T. Novel three-dimensional zinc oxide superstructures for high dielectric constant polymer composites capable of withstanding high electric field. *J Phys Chem C* 2012;116:24887–95. https://doi.org/10.1021/jp3088644.

492. Yu K, Niu Y, Xiang F, Zhou Y, Bai Y, Wang H. Enhanced electric breakdown strength and high energy density of barium titanate filled polymer nanocomposites. *J Appl Phys* 2013;114:174107. https://doi.org/10.1063/1.4829671.

493. Luo H, Zhang D, Jiang C, Yuan X, Chen C, Zhou K. Improved dielectric properties and energy storage density of poly(vinylidene fluoride- co -hexafluoropropylene) nanocomposite with hydantoin epoxy resin coated BaTiO$_3$. *ACS Appl Mater Interfaces* 2015;7:8061–9. https://doi.org/10.1021/acsami.5b00555.

494. Niu Y, Yu K, Bai Y, Wang H. Enhanced dielectric performance of $BaTiO_3$/PVDF composites prepared by modified process for energy storage applications. *IEEE Trans Ultrason Ferroelectr Freq Control* 2015;62:108–15. https://doi.org/10.1109/TUFFC.2014.006666.
495. Yang K, Huang X, Huang Y, Xie L, Jiang P. Fluoro-polymer@ $BaTiO_3$ hybrid nanoparticles prepared via RAFT polymerization: Toward ferroelectric polymer nanocomposites with high dielectric constant and low dielectric loss for energy storage application. *Chem Mater* 2013;25:2327–38. https://doi.org/10.1021/cm4010486.
496. Yu K, Wang H, Zhou Y, Bai Y, Niu Y. Enhanced dielectric properties of $BaTiO_3$/poly(vinylidene fluoride) nanocomposites for energy storage applications. *J Appl Phys* 2013;113:034105. https://doi.org/10.1063/1.4776740.
497. Yu K, Niu Y, Zhou Y, Bai Y, Wang H. Nanocomposites of surface-modified $BaTiO_3$ nanoparticles filled ferroelectric polymer with enhanced energy density. *J Am Ceram Soc* 2013;96:2519–24. https://doi.org/10.1111/jace.12338.
498. Tang H, Lin Y, Andrews C, Sodano HA. Nanocomposites with increased energy density through high aspect ratio PZT nanowires. *Nanotechnology* 2011;22:015702. https://doi.org/10.1088/0957-4484/22/1/015702.
499. Dang Z-M, Wang H-Y, Peng B, Nan C-W. Effect of $BaTiO_3$ size on dielectric property of $BaTiO_3$/PVDF composites. *J Electroceramics* 2008;21:381–4. https://doi.org/10.1007/s10832-007-9201-8.
500. Dang Z-M, Zhou T, Yao S-H, Yuan J-K, Zha J-W, Song H-T, et al. Advanced calcium copper titanate/polyimide functional hybrid films with high dielectric permittivity. *Adv Mater* 2009;21:2077–82. https://doi.org/10.1002/adma.200803427.
501. Apostolova IN, Apostolov AT, Golrokh Bahoosh S, Wesselinowa JM. Origin of ferromagnetism in transition metal doped $BaTiO_3$. *J Appl Phys* 2013;113:203904. https://doi.org/10.1063/1.4807412.
502. Ju L, Sabergharesou T, Stampecoskie KG, Hegde M, Wang T, Combe NA, et al. Interplay between size, composition, and phase transition of nanocrystalline Cr^{3+}-doped $BaTiO_3$ as a path to multiferroism in perovskite-type oxides. *J Am Chem Soc* 2012;134:1136–46. https://doi.org/10.1021/ja2091678.
503. Shah J, Kotnala RK. Induced magnetism and magnetoelectric coupling in ferroelectric $BaTiO_3$ by Cr-doping synthesized by a facile chemical route. *J Mater Chem A* 2013;1:8601. https://doi.org/10.1039/c3ta11845b.
504. Rajan S, Gazzali PMM, Chandrasekaran G. Electrical and magnetic phase transition studies of Fe and Mn co-doped $BaTiO_3$. *J Alloys Compd* 2016;656:98–109. https://doi.org/10.1016/j.jallcom.2015.09.199.
505. Gao X, Li L, Jian J, Wang H, Fan M, Huang J, et al. Vertically aligned nanocomposite $BaTiO_3$: $YMnO_3$ thin films with room temperature multiferroic properties toward nanoscale memory devices. *ACS Appl Nano Mater* 2018;1:2509–14. https://doi.org/10.1021/acsanm.8b00614.
506. Rajamani A, Dionne GF, Bono D, Ross CA. Faraday rotation, ferromagnetism, and optical properties in Fe-doped $BaTiO_3$. *J Appl Phys* 2005;98:063907. https://doi.org/10.1063/1.2060945.
507. Lin Y-H, Yuan J, Zhang S, Zhang Y, Liu J, Wang Y, et al. Multiferroic behavior observed in highly orientated Mn-doped $BaTiO_3$ thin films. *Appl Phys Lett* 2009;95:033105. https://doi.org/10.1063/1.3182793.

508. Coey JMD, Venkatesan M, Fitzgerald CB. Donor impurity band exchange in dilute ferromagnetic oxides. *Nat Mater* 2005;4:173–9. https://doi.org/10.1038/nmat1310.

509. Shuai Y, Zhou S, Bürger D, Reuther H, Skorupa I, John V, et al. Decisive role of oxygen vacancy in ferroelectric versus ferromagnetic Mn-doped $BaTiO_3$ thin films. *J Appl Phys* 2011;109:084105. https://doi.org/10.1063/1.3576125.

510. Mangalam RVK, Ray N, Waghmare U V., Sundaresan A, Rao CNR. Multiferroic properties of nanocrystalline $BaTiO_3$. *Solid State Commun* 2009;149:1–5. https://doi.org/10.1016/j.ssc.2008.10.023.

511. Wang M, Tan G-L, Zhang Q. Multiferroic properties of nanocrystalline $PbTiO_3$ ceramics. *J Am Ceram Soc* 2010;93:2151–4. https://doi.org/10.1111/j.1551-2916.2010.03691.x.

512. Costanzo T, McCracken J, Rotaru A, Caruntu G. Quasi-monodisperse transition-metal-doped $BaTiO_3$ (M=Cr, Mn, Fe, Co) colloidal nanocrystals with multiferroic properties. *ACS Appl Nano Mater* 2018;1:4863–74. https://doi.org/10.1021/acsanm.8b01036.

513. Smith MB, Page K, Siegrist T, Redmond PL, Walter EC, Seshadri R, et al. Crystal structure and the paraelectric-to-ferroelectric phase transition of nanoscale $BaTiO_3$. *J Am Chem Soc* 2008;130:6955–63. https://doi.org/10.1021/ja0758436.

514. Rodriguez BJ, Callahan C, Kalinin S V, Proksch R. Dual-frequency resonance-tracking atomic force microscopy. *Nanotechnology* 2007;18:475504. https://doi.org/10.1088/0957-4484/18/47/475504.

515. Jesse S, Baddorf AP, Kalinin S V. Switching spectroscopy piezoresponse force microscopy of ferroelectric materials. *Appl Phys Lett* 2006;88:062908. https://doi.org/10.1063/1.2172216.

516. Yang L, Qiu H, Pan L, Guo Z, Xu M, Yin J, et al. Magnetic properties of $BaTiO_3$ and $BaTi_{1-x}M_xO_3$ (M=Co, Fe) nanocrystals by hydrothermal method. *J Magn Magn Mater* 2014;350:1–5. https://doi.org/10.1016/j.jmmm.2013.09.036.

517. Bahoosh SG, Trimper S, Wesselinowa JM. Origin of ferromagnetism in $BaTiO_3$ nanoparticles. *Phys Status Solidi - Rapid Res Lett* 2011;5:382–4. https://doi.org/10.1002/pssr.201105419.

518. Sundaresan A, Rao CNR. Implications and consequences of ferromagnetism universally exhibited by inorganic nanoparticles. *Solid State Commun* 2009;149:1197–200. https://doi.org/10.1016/j.ssc.2009.04.028.

519. Li X, Lv Z, Zhu H. Carbon/silicon heterojunction solar cells: State of the art and prospects. *Adv Mater* 2015;27:6549–74. https://doi.org/10.1002/adma.201502999.

520. Falcão AF de O. Wave energy utilization: A review of the technologies. *Renew Sustain Energy Rev* 2010;14:899–918. https://doi.org/10.1016/j.rser.2009.11.003.

521. Park K-I, Lee M, Liu Y, Moon S, Hwang G-T, Zhu G, et al. Flexible nanocomposite generator made of $BaTiO_3$ nanoparticles and graphitic carbons. *Adv Mater* 2012;24:2999–3004. https://doi.org/10.1002/adma.201200105.

522. Lee KY, Kim D, Lee J-H, Kim TY, Gupta MK, Kim S-W. Unidirectional high-power generation via stress-induced dipole alignment from $ZnSnO_3$ nanocubes/polymer hybrid piezoelectric nanogenerator. *Adv Funct Mater* 2014;24:37–43. https://doi.org/10.1002/adfm.201301379.

523. Hu Y, Wang ZL. Recent progress in piezoelectric nanogenerators as a sustainable power source in self-powered systems and active sensors. *Nano Energy* 2015;14:3–14. https://doi.org/10.1016/j.nanoen.2014.11.038.

524. Park K-I, Son JH, Hwang G-T, Jeong CK, Ryu J, Koo M, et al. Nanogenerators: Highly-efficient, flexible piezoelectric PZT thin film nanogenerator on plastic substrates (Adv. Mater. 16/2014). *Adv Mater* 2014;26:2450–2450. https://doi.org/10.1002/adma.201470103.

525. Li Z, Wang ZL. Air/liquid-pressure and heartbeat-driven flexible fiber nanogenerators as a micro/nano-power source or diagnostic sensor. *Adv Mater* 2011;23:84–9. https://doi.org/10.1002/adma.201003161.

526. Xue X, Wang S, Guo W, Zhang Y, Wang ZL. Hybridizing energy conversion and storage in a mechanical-to-electrochemical process for self-charging power cell. *Nano Lett* 2012;12:5048–54. https://doi.org/10.1021/nl302879t.

527. Gupta MK, Lee J-H, Lee KY, Kim S-W. Two-dimensional vanadium-doped ZnO nanosheet-based flexible direct current nanogenerator. *ACS Nano* 2013;7:8932–9. https://doi.org/10.1021/nn403428m.

528. Jung JH, Chen C-Y, Yun BK, Lee N, Zhou Y, Jo W, et al. Lead-free KNbO$_3$ ferroelectric nanorod based flexible nanogenerators and capacitors. *Nanotechnology* 2012;23:375401. https://doi.org/10.1088/0957-4484/23/37/375401.

529. Wu JM, Xu C, Zhang Y, Yang Y, Zhou Y, Wang ZL. Flexible and transparent nanogenerators based on a composite of lead-free ZnSnO$_3$ triangular-belts. *Adv Mater* 2012;24:6094–9. https://doi.org/10.1002/adma.201202445.

530. Persano L, Dagdeviren C, Su Y, Zhang Y, Girardo S, Pisignano D, et al. High performance piezoelectric devices based on aligned arrays of nanofibers of poly(vinylidenefluoride-co-trifluoroethylene). *Nat Commun* 2013;4:1633. https://doi.org/10.1038/ncomms2639.

531. Yan J, Jeong YG. High performance flexible piezoelectric nanogenerators based on BaTiO$_3$ nanofibers in different alignment modes. *ACS Appl Mater Interfaces* 2016;8:15700–9. https://doi.org/10.1021/acsami.6b02177.

532. Chen LF, Hong YP, Chen XJ, Wu QL, Huang QJ, Luo XT. Preparation and properties of polymer matrix piezoelectric composites containing aligned BaTiO$_3$ whiskers. *J Mater Sci* 2004;39:2997–3001. https://doi.org/10.1023/B:JMSC.0000025825.04281.89.

533. Carponcin D, Dantras E, Dandurand J, Aridon G, Levallois F, Cadiergues L, et al. Electrical and piezoelectric behavior of polyamide/PZT/CNT multifunctional nanocomposites. *Adv Eng Mater* 2014;16:1018–25. https://doi.org/10.1002/adem.201300519.

534. Ni X, Wang F, Lin A, Xu Q, Yang Z, Qin Y. Flexible nanogenerator based on single BaTiO$_3$ nanowire. *Sci Adv Mater* 2013;5:1781–7. https://doi.org/10.1166/sam.2013.1629.

535. Zhang M, Gao T, Wang J, Liao J, Qiu Y, Xue H, et al. Single BaTiO$_3$ nanowires-polymer fiber based nanogenerator. *Nano Energy* 2015;11:510–7. https://doi.org/10.1016/j.nanoen.2014.11.028.

536. Gao T, Liao J, Wang J, Qiu Y, Yang Q, Zhang M, et al. Highly oriented BaTiO$_3$ film self-assembled using an interfacial strategy and its application as a flexible piezoelectric generator for wind energy harvesting. *J Mater Chem A* 2015;3:9965–71. https://doi.org/10.1039/C5TA01079A.

537. Park K-I, Xu S, Liu Y, Hwang G-T, Kang S-JL, Wang ZL, et al. Piezoelectric BaTiO$_3$ thin film nanogenerator on plastic substrates. *Nano Lett* 2010;10:4939–43. https://doi.org/10.1021/nl102959k.

538. Lin Z-H, Yang Y, Wu JM, Liu Y, Zhang F, Wang ZL. BaTiO$_3$ nanotubes-based flexible and transparent nanogenerators. *J Phys Chem Lett* 2012;3:3599–604. https://doi.org/10.1021/jz301805f.

539. Jeong CK, Kim I, Park K-I, Oh MH, Paik H, Hwang G-T, et al. Virus-directed design of a flexible BaTiO$_3$ nanogenerator. *ACS Nano* 2013;7:11016–25. https://doi.org/10.1021/nn404659d.

540. Shin S-H, Kim Y-H, Lee MH, Jung J-Y, Nah J. Hemispherically aggregated BaTiO$_3$ nanoparticle composite thin film for high-performance flexible piezoelectric nanogenerator. *ACS Nano* 2014;8:2766–73. https://doi.org/10.1021/nn406481k.

541. Wu Y-F, Nien Y-T, Wang Y-J, Chen I-G. Enhancement of photoluminescence and color purity of CaTiO$_3$: Eu phosphor by Li doping. *J Am Ceram Soc* 2012;95:1360–6. https://doi.org/10.1111/j.1551-2916.2011.04967.x.

542. Milanez J, de Figueiredo AT, de Lazaro S, Longo VM, Erlo R, Mastelaro VR, et al. The role of oxygen vacancy in the photoluminescence property at room temperature of the CaTiO$_3$. *J Appl Phys* 2009;106:043526. https://doi.org/10.1063/1.3190524.

543. Chen Y, Patel S, Ye Y, Shaw DT, Guo L. Field emission from aligned high-density graphitic nanofibers. *Appl Phys Lett* 1998;73:2119–21. https://doi.org/10.1063/1.122397.

544. Takashima H, Ueda K, Itoh M. Red photoluminescence in praseodymium-doped titanate perovskite films epitaxially grown by pulsed laser deposition. *Appl Phys Lett* 2006;89:261915. https://doi.org/10.1063/1.2424438.

545. Kyômen T, Sakamoto R, Sakamoto N, Kunugi S, Itoh M. Photoluminescence properties of Pr-doped (Ca,Sr,Ba)TiO$_3$. *Chem Mater* 2005;17:3200–4. https://doi.org/10.1021/cm0403715.

546. Tiwari A, Dhoble SJ. Tunable lanthanide/transition metal ion-doped novel phosphors for possible application in w-LEDs: A review. *Luminescence* 2019;35:4–33. https://doi.org/10.1002/bio.3712.

547. Nair GB, Swart HC, Dhoble SJ. A review on the advancements in phosphor-converted light emitting diodes (pc-LEDs): Phosphor synthesis, device fabrication and characterization. *Prog Mater Sci* 2019:100622. https://doi.org/10.1016/j.pmatsci.2019.100622.

548. Som S, Kunti AK, Kumar V, Kumar V, Dutta S, Chowdhury M, et al. Defect correlated fluorescent quenching and electron phonon coupling in the spectral transition of Eu^{3+} in CaTiO$_3$ for red emission in display application. *J Appl Phys* 2014;115:193101. https://doi.org/10.1063/1.4876316.

549. Parchur AK, Ningthoujam RS. Preparation and structure refinement of Eu^{3+} doped CaMoO$_4$ nanoparticles. *Dalt Trans* 2011;40:7590. https://doi.org/10.1039/c1dt10327j.

550. Lu D-Y, Sun X-Y, Toda M. Electron spin resonance investigations and compensation mechanism of europium-doped barium titanate ceramics. *Jpn J Appl Phys* 2006;45:8782–8. https://doi.org/10.1143/JJAP.45.8782.

551. Schmechel R, Kennedy M, von Seggern H, Winkler H, Kolbe M, Fischer RA, et al. Luminescence properties of nanocrystalline Y$_2$O$_3$:Eu^{3+} in different host materials. *J Appl Phys* 2001;89:1679. https://doi.org/10.1063/1.1333033.

552. Sun Y, Qi L, Lee M, Lee B., Samuels W., Exarhos G. Photoluminescent properties of Y$_2$O$_3$:Eu^{3+} phosphors prepared via urea precipitation in non-aqueous solution. *J Lumin* 2004;109:85–91. https://doi.org/10.1016/j.jlumin.2004.01.085.

553. Li Y-C, Chang Y-H, Lin Y-F, Chang Y-S, Lin Y-J. Synthesis and luminescent properties of Ln^{3+} (Eu^{3+}, Sm^{3+}, Dy^{3+})-doped lanthanum aluminum germanate LaAlGe$_2$O$_7$ phosphors. *J Alloys Compd* 2007;439:367–75. https://doi.org/10.1016/j.jallcom.2006.08.269.

554. Som S, Sharma SK. Eu^{3+} /Tb^{3+} -codoped Y$_2$O$_3$ nanophosphors: Rietveld refinement, bandgap and photoluminescence optimization. *J Phys D Appl Phys* 2012;45:415102. https://doi.org/10.1088/0022-3727/45/41/415102.

555. Boutinaud P, Sarakha L, Cavalli E, Bettinelli M, Dorenbos P, Mahiou R. About red afterglow in Pr^{3+} doped titanate perovskites. *J Phys D Appl Phys* 2009;42:045106. https://doi.org/10.1088/0022-3727/42/4/045106.

556. van Dijk JMF, Schuurmans MFH. On the nonradiative and radiative decay rates and a modified exponential energy gap law for 4 f –4 f transitions in rare-earth ions. *J Chem Phys* 1983;78:5317–23. https://doi.org/10.1063/1.445485.

557. Liu B, Gu M, Liu X, Han K, Huang S, Ni C, et al. Enhanced luminescence through ion-doping-induced higher energy phonons in GdTaO$_4$: Eu^{3+} phosphor. *Appl Phys Lett* 2009;94:061906. https://doi.org/10.1063/1.3079413.

558. Dutta S, Som S, Sharma SK. Luminescence and photometric characterization of K$^+$ compensated CaMoO$_4$: Dy^{3+} nanophosphors. *Dalt Trans* 2013;42:9654. https://doi.org/10.1039/c3dt50780g.

559. Dexter DL. A theory of sensitized luminescence in solids. *J Chem Phys* 1953;21:836–50. https://doi.org/10.1063/1.1699044.

560. Blasse G. Energy transfer between inequivalent Eu^{2+} ions. *J Solid State Chem* 1986;62:207–11. https://doi.org/10.1016/0022-4596(86)90233-1.

561. Dexter DL, Schulman JH. Theory of concentration quenching in inorganic phosphors. *J Chem Phys* 1954;22:1063–70. https://doi.org/10.1063/1.1740265.

562. Downing E, Hesselink L, Ralston J, Macfarlane R. A three-color, solid-state, three-dimensional display. *Science* 1996;273:1185–9. https://doi.org/10.1126/science.273.5279.1185.

563. Wolfbeis OS, Dürkop A, Wu M, Lin Z. A europium-ion-based luminescent sensing probe for hydrogen peroxide. *Angew Chemie Int Ed* 2002;41:4495–8. https://doi.org/10.1002/1521-3773(20021202)41:23<4495::AID-ANIE4495>3.0.CO;2-I.

564. Akizuki N, Aota S, Mouri S, Matsuda K, Miyauchi Y. Efficient near-infrared upconversion photoluminescence in carbon nanotubes. *Nat Commun* 2015;6:8920. https://doi.org/10.1038/ncomms9920.

565. Wang F, Liu X. Recent advances in the chemistry of lanthanide-doped upconversion nanocrystals. *Chem Soc Rev* 2009;38:976. https://doi.org/10.1039/b809132n.

566. Auzel F. Upconversion and anti-stokes processes with f and d ions in solids. *Chem Rev* 2004;104:139–74. https://doi.org/10.1021/cr020357g.

567. Li B, Zhang S, Zhou X, Chen Z, Wang S. Microstructures and dielectric properties of Y/Zn codoped BaTiO$_3$ ceramics. *J Mater Sci* 2007;42:5223–8. https://doi.org/10.1007/s10853-006-0604-8.

568. Mahata MK, Koppe T, Mondal T, Brüsewitz C, Kumar K, Kumar Rai V, et al. Incorporation of Zn^{2+} ions into BaTiO$_3$: Er^{3+} /Yb^{3+} nanophosphor: An effective way to enhance upconversion, defect luminescence and temperature sensing. *Phys Chem Chem Phys* 2015;17:20741–53. https://doi.org/10.1039/C5CP01874A.

569. Jo SK, Park JS, Han YH. Effects of multi-doping of rare-earth oxides on the microstructure and dielectric properties of BaTiO$_3$. *J Alloys Compd* 2010;501:259–64. https://doi.org/10.1016/j.jallcom.2010.04.085.

570. Supasai T, Dangtip S, Learngarunsri P, Boonyopakorn N, Wisitsoraat A, Hodak SK. Influence of temperature annealing on optical properties of SrTiO$_3$/BaTiO$_3$ multilayered films on indium tin oxide. *Appl Surf Sci* 2010;256:4462–7. https://doi.org/10.1016/j.apsusc.2010.01.072.

571. Yang L, Li L, Zhao M, Li G. Size-induced variations in bulk/surface structures and their impact on photoluminescence properties of $GdVO_4$: Eu^{3+} nanoparticles. *Phys Chem Chem Phys* 2012;14:9956. https://doi.org/10.1039/c2cp41136a.

572. Cheng Q, Sui J, Cai W. Enhanced upconversion emission in Yb^{3+} and Er^{3+} codoped $NaGdF_4$ nanocrystals by introducing Li^+ ions. *Nanoscale* 2012;4:779–84. https://doi.org/10.1039/C1NR11365H.

573. Glais E, Đorđević V, Papan J, Viana B, Dramićanin MD. $MgTiO_3$: Mn^{4+} a multi-reading temperature nanoprobe. *RSC Adv* 2018;8:18341–6. https://doi.org/10.1039/C8RA02482K.

574. Nikolić MG, Antić Ž, Ćulubrk S, Nedeljković JM, Dramićanin MD. Temperature sensing with Eu^{3+} doped TiO_2 nanoparticles. *Sensors Actuators B Chem* 2014;201:46–50. https://doi.org/10.1016/j.snb.2014.04.108.

575. Savchuk OA, Carvajal JJ, Cascales C, Aguiló M, Díaz F. Benefits of silica core–shell structures on the temperature sensing properties of Er, Yb: $GdVO_4$ upconversion nanoparticles. *ACS Appl Mater Interfaces* 2016;8:7266–73. https://doi.org/10.1021/acsami.6b01371.

576. Balabhadra S, Debasu ML, Brites CDS, Nunes LAO, Malta OL, Rocha J, et al. Boosting the sensitivity of Nd^{3+} -based luminescent nanothermometers. *Nanoscale* 2015;7:17261–7. https://doi.org/10.1039/C5NR05631D.

577. Cortelletti P, Facciotti C, Cantarelli IX, Canton P, Quintanilla M, Vetrone F, et al. Nd^{3+} activated CaF_2 NPs as colloidal nanothermometers in the biological window. *Opt Mater (Amst)* 2017;68:29–34. https://doi.org/10.1016/j.optmat.2016.11.019.

578. Back M, Trave E, Ueda J, Tanabe S. Ratiometric optical thermometer based on dual near-infrared emission in Cr^{3+} -doped bismuth-based gallate host. *Chem Mater* 2016;28:8347–56. https://doi.org/10.1021/acs.chemmater.6b03625.

579. Marciniak L, Bednarkiewicz A, Kowalska D, Strek W. A new generation of highly sensitive luminescent thermometers operating in the optical window of biological tissues. *J Mater Chem C* 2016;4:5559–63. https://doi.org/10.1039/C6TC01484D.

580. Li F, Wang F, Hu X, Zheng B, Du J, Xiao D. A long-persistent phosphorescent chemosensor for the detection of TNP based on $CaTiO_3$: $Pr^{3+}@SiO_2$ photoluminescence materials. *RSC Adv* 2018;8:16603–10. https://doi.org/10.1039/C8RA02665C.

581. Xiong C, Pernice WHP, Tang HX. Low-loss, silicon integrated, aluminum nitride photonic circuits and their use for electro-optic signal processing. *Nano Lett* 2012;12:3562–8. https://doi.org/10.1021/nl3011885.

582. Raju SH, Sudhakar BM, Reddy BS, Dhoble SJ, Thyagarajan K, Raju CN. Synthesis, photoluminescence and thermoluminescence properties of Sm^{3+} and Dy^{3+} ions doped barium gadolinium titanate ceramics. *Ferroelectr Lett Sect* 2014;41:9–19. https://doi.org/10.1080/07315171.2014.908680.

583. Lin S, Vetter RJ, Ziemer PL. Thermoluminescence and microwave induced thermoluminescence fading of rare-earth-doped barium titanate ceramics. *Radiat Eff* 1978;38:67–71. https://doi.org/10.1080/00337577808233210.

584. Horchidan N, Padurariu L, Ciomaga CE, Curecheriu L, Airimioaei M, Doroftei F, et al. Room temperature phase superposition as origin of enhanced functional properties in $BaTiO_3$- based ceramics. *J Eur Ceram Soc* 2019. https://doi.org/10.1016/j.jeurceramsoc.2019.11.088.

585. Fasasi AY, Balogun FA, Fasasi MK, Ogunleye PO, Mokobia CE, Inyang EP. Thermoluminescence properties of barium titanate prepared by solid-state reaction. *Sensors Actuators, A Phys* 2007;135:598–604. https://doi.org/10.1016/j.sna.2006.07.029.

586. Tiwari N, Dubey V, Dewangan J, Jain N. Near UV-blue emission from cerium doped zirconium dioxide phosphor for display and sensing applications. *J Disp Technol* 2016;12:933–7. https://doi.org/10.1109/JDT.2016.2544881.

587. Lu DY, Peng YY. Dielectric properties and exploration of self-compensation mode of Tb in $BaTiO_3$ ceramics. *J Ceram Soc Japan* 2016;124:455–9. https://doi.org/10.2109/jcersj2.15292.

588. Pan E, Bai G, Lei L, Zhang J, Xu S. The electrical enhancement and reversible manipulation of near-infrared luminescence in Nd doped ferroelectric nano-composites for optical switches. *J Mater Chem C* 2019;7:4320–5. https://doi.org/10.1039/c9tc00286c.

589. Awan IT, Rivera VAG, Nogueira IC, Pereira-da-Silva MA, Li MS, et al. Growth process and grain boundary defects in Er doped $BaTiO_3$ processed by EB-PVD: A study by XRD, FTIR, SEM and AFM. *Appl Surf Sci* 2019;493:982–93. https://doi.org/10.1016/j.apsusc.2019.07.003.

590. Kadam AR, Mishra GC, Dhoble SJ. Thermoluminescence study of Eu^{3+} doped $Na_2Sr_2Al_2PO_4C_{19}$ phosphor via doping of singly, doubly and triply ionized ions. *Ceram Int* 2020;46(1):132–55. https://doi.org/10.1016/j.ceramint.2019.08.242.

591. Daniel DJ, Kim HJ, Kim S, Kothan S, Kaewkhao J. Trap level analysis of Ce^{3+} and Sm^{3+} in $Li_6Y(BO_3)_3$. *Ceram Int* 2019;45:11893–8. https://doi.org/10.1016/j.ceramint.2019.03.072.

592. Dubey V, Kaur J, Dubey N, Pandey MK, Suryanarayana NS, Murthy KVR. Kinetic and TL glow curve analysis of UV-, β- and γ-irradiated natural limestone collected from Chunkatta mines. *Radiat Eff Defects Solids* 2017;172:866–77. https://doi.org/10.1080/10420150.2017.1417410.

593. Singh R, Kaur J, Bose P, Shrivastava R, Dubey V, Parganiha Y. Intense visible light emission from dysprosium (Dy^{3+}) doped barium titanate ($BaTiO_3$) phosphor and its thermoluminescence study. *J Mater Sci Mater Electron* 2017;28:13690–7. https://doi.org/10.1007/s10854-017-7212-z.

594. Sun Q, Gu Q, Zhu K, Wang J, Qiu J. Stabilized temperature-dependent dielectric properties of Dy-doped $BaTiO_3$ ceramics derived from sol-hydrothermally synthesized nanopowders. *Ceram Int* 2016;42:3170–6. https://doi.org/10.1016/j.ceramint.2015.10.107.

595. Liu Q, Guo J, Fan M, Zhang Q, Liu S, Wong KL, Liu Z, Wei B. Fast synthesis of Dy^{3+} and Tm^{3+} co-doped double perovskite $NaLaMgWO_6$: A thermally stable singlephase white-emitting phosphor for WLEDs. *J Mater Chem* 2013;1:3777. https://doi.org/10.1039/b000000x.

596. Song E, Zhao W, Zhou G, Dou X, Yi C, Zhou M. Luminescence properties of red phosphors $Ca_{10}Li(PO_4)_7$: Eu^{3+}. *J Rare Earths* 2011;29:440–3. https://doi.org/10.1016/S1002-0721(10)60476-0.

597. Wang W, Tao M, Liu Y, Wei Y, Xing G, Dang P, et al. Photoluminescence control of UCr_4C_4-type phosphors with superior luminous efficiency and high color purity via controlling site selection of Eu^{2+} activators. *Chem Mater* 2019. https://doi.org/10.1021/acs.chemmater.9b04089.

598. Kadam AR, Dhoble SJ, Yadav RS. Effect of singly, doubly and triply ionized ions on downconversion photoluminescence in Eu^{3+} doped $Na_2Sr_2Al_2PO_4C_{19}$ phosphor: A comparative study. *Ceram Int* 2020;46(3):3264–74. https://doi.org/10.3786/nml.v4i2.p78-82.

599. Kumari A, Rai VK, Kumar K. Yellow-orange upconversion emission in Eu^{3+}-Yb^{3+}codoped $BaTiO_3$ phosphor. *Spectrochim Acta - Part A Mol Biomol Spectrosc* 2014;127:98–101. https://doi.org/10.1016/j.saa.2014.02.023.

600. Peng Z, Wu D, Liang P, Zhu J, Zhou X, Chao X, et al. Understanding the ultra-high dielectric permittivity response in titanium dioxide ceramics. *Ceram Int* 2020;46:2545–51. https://doi.org/10.1016/j.ceramint.2019.09.109.

601. Spanier JE, Kolpak AM, Urban JJ, Grinberg I, Ouyang L, Yun WS, et al. Ferroelectric phase transition in individual single-crystalline $BaTiO_3$ nanowires. *Nano Lett* 2006;6:735–9. https://doi.org/10.1021/nl052538e.

602. Tian G, Song J, Liu J, Qi S, Wu D. Enhanced dielectric permittivity and thermal stability of graphene-polyimide nanohybrid films. *Soft Mater* 2014;12:290–6. https://doi.org/10.1080/1539445X.2014.902852.

603. Zhang X, Huan Y, Zhu Y, Tian H, Li K, Hao Y, et al. Enhanced photocatalytic activity by the combined influence of ferroelectric domain and Au nanoparticles for $BaTiO_3$ fibers. *Nano* 2018;13:1–10. https://doi.org/10.1142/S1793292018501497.

604. Kim SD, Hwang GT, Song K, Jeong CK, Park K Il, Jang J, et al. Inverse size-dependence of piezoelectricity in single $BaTiO_3$ nanoparticles. *Nano Energy* 2019;58:78–84. https://doi.org/10.1016/j.nanoen.2018.12.096.

605. Deng X, Wang X, Wen H, Kang A, Gui Z, Li L. Phase transitions in nanocrystalline barium titanate ceramics prepared by spark plasma sintering. *J Am Ceram Soc* 2006;89:1059–64. https://doi.org/10.1111/j.1551-2916.2005.00836.x.

606. Pan JH, Lee WI. Preparation of highly ordered cubic mesoporous WO_3/TiO_2 films and their photocatalytic properties. *Chem Mater* 2006;18:847–53. https://doi.org/10.1021/cm0522782.

607. Wen B, Liu C, Liu Y. Depositional characteristics of metal coating on single-crystal TiO_2 nanowires. *J Phys Chem B* 2005;109:12372–5. https://doi.org/10.1021/jp050934f.

608. Kulkarni SA, Mhaisalkar SG, Mathews N, Boix PP. Perovskite nanoparticles: Synthesis, properties, and novel applications in photovoltaics and LEDs. *Small Methods* 2019;3:1–16. https://doi.org/10.1002/smtd.201800231.

609. Yuan Q, Li G, Yao FZ, Cheng SD, Wang Y, Ma R, et al. Simultaneously achieved temperature-insensitive high energy density and efficiency in domain engineered $BaTiO_3$-$Bi(Mg_{0.5}Zr_{0.5})O_3$ lead-free relaxor ferroelectrics. *Nano Energy* 2018;52:203–10. https://doi.org/10.1016/j.nanoen.2018.07.055.

610. Tsuji K, Ndayishimiye A, Lowum S, Floyd R, Wang K, Wetherington M, et al. Single step densification of high permittivity $BaTiO_3$ ceramics at 300°C. *J Eur Ceram Soc* 2019. https://doi.org/10.1016/j.jeurceramsoc.2019.12.022.

611. Li Y, Li W, Du G, Chen N. Low temperature preparation of $CaCu_3Ti_4O_{12}$ ceramics with high permittivity and low dielectric loss. *Ceram Int* 2017;43:9178–83. https://doi.org/10.1016/j.ceramint.2017.04.069.

612. Lin Y, Wang J, Jiang L, Chen Y, Nan CW. High permittivity Li and Al doped NiO ceramics. *Appl Phys Lett* 2004;85:5664–6. https://doi.org/10.1063/1.1827937.

613. Wang Z, Wen YF, Li HJ, Fang MR, Wang C, Pu YP. Excellent stability and low dielectric loss of $Ba(Fe_{0.5}Nb_{0.5})O_3$ synthesized by a solution precipitation method. *J Alloys Compd* 2016;656:431–8. https://doi.org/10.1016/j.jallcom.2015.09.169.

614. Tkach A, Okhay O, Almeida A, Vilarinho PM. Giant dielectric permittivity and high tunability in Y-doped SrTiO$_3$ ceramics tailored by sintering atmosphere. *Acta Mater* 2017;130:249–60. https://doi.org/10.1016/j.actamat.2017.03.051.

615. Hu W, Liu Y, Withers RL, Frankcombe TJ, Norén L, Snashall A, et al. Electron-pinned defect-dipoles for high-performance colossal permittivity materials. *Nat Mater* 2013;12:821–6. https://doi.org/10.1038/nmat3691.

616. Wei X, Jie W, Yang Z, Zheng F, Zeng H, Liu Y, et al. Colossal permittivity properties of Zn, Nb co-doped TiO$_2$ with different phase structures. *J Mater Chem C* 2015;3:11005–10. https://doi.org/10.1039/c5tc02578h.

617. Li Z, Luo X, Wu W, Wu J. Niobium and divalent-modified titanium dioxide ceramics: Colossal permittivity and composition design. *J Am Ceram Soc* 2017;100:3004–12. https://doi.org/10.1111/jace.14850.

618. Yang C, Tse MY, Wei X, Hao J. Colossal permittivity of (Mg+Nb) co-doped TiO$_2$ ceramics with low dielectric loss. *J Mater Chem C* 2017;5:5170–5. https://doi.org/10.1039/c7tc01020f.

619. Tse MY, Wei X, Hao J. High-performance colossal permittivity materials of (Nb+Er) co-doped TiO$_2$ for large capacitors and high-energy-density storage devices. *Phys Chem Chem Phys* 2016;18:24270–7. https://doi.org/10.1039/c6cp02236g.

620. Song Y, Wang X, Zhang X, Qi X, Liu Z, Zhang L, et al. Colossal dielectric permittivity in (Al+Nb) co-doped rutile SnO$_2$ ceramics with low loss at room temperature. *Appl Phys Lett* 2016;109:1–6. https://doi.org/10.1063/1.4964121.

621. Wang X, Zhang B, Shen G, Sun L, Hu Y, Shi L, et al. Colossal permittivity and impedance analysis of tantalum and samarium co-doped TiO$_2$ ceramics. *Ceram Int* 2017;43:13349–55. https://doi.org/10.1016/j.ceramint.2017.07.034.

622. Dong W, Hu W, Berlie A, Lau K, Chen H, Withers RL, et al. Colossal dielectric behavior of Ga+Nb co-doped rutile TiO$_2$. *ACS Appl Mater Interfaces* 2015;7:25321–5. https://doi.org/10.1021/acsami.5b07467.

623. Song Y, Liu P, Zhao X, Guo B, Cui X. Dielectric properties of (Bi$_{0.5}$Nb$_{0.5}$)xTi$_{1-x}$O$_2$ ceramics with colossal permittivity. *J Alloys Compd* 2017;722:676–82. https://doi.org/10.1016/j.jallcom.2017.06.177.

624. Fan J, Leng S, Cao Z, He W, Gao Y, Liu J, et al. Colossal permittivity of Sb and Ga co-doped rutile TiO$_2$ ceramics. *Ceram Int* 2019;45:1001–10. https://doi.org/10.1016/j.ceramint.2018.09.279.

625. Wang Z, Chen H, Wang T, Xiao Y, Nian W, Fan J. Enhanced relative permittivity in niobium and europium co-doped TiO$_2$ ceramics. *J Eur Ceram Soc* 2018;38:3847–52. https://doi.org/10.1016/j.jeurceramsoc.2018.04.026.

626. Chen J, Park NG. Inorganic hole transporting materials for stable and high efficiency perovskite solar cells. *J Phys Chem C* 2018;122:14039–63. https://doi.org/10.1021/acs.jpcc.8b01177.

627. Lin L, Jiang L, Li P, Fan B, Qiu Y. A modeled perovskite solar cell structure with a Cu 2 O hole-transporting layer enabling over 20% efficiency by low-cost low-temperature processing. *J Phys Chem Solids* 2019;124:205–11. https://doi.org/10.1016/j.jpcs.2018.09.024.

628. Jeon NJ, Na H, Jung EH, Yang TY, Lee YG, Kim G, et al. A fluorene-terminated hole-transporting material for highly efficient and stable perovskite solar cells. *Nat Energy* 2018;3:682–9. https://doi.org/10.1038/s41560-018-0200-6.

629. T M, A K, K T, Y S. Organometal halide perovskites as visible-light sensitizers for photovoltaic cells. *J Am Chem Soc* 2009;131:6050–1.

630. Borchert J, Milot RL, Patel JB, Davies CL, Wright AD, Martínez Maestro L, et al. Large-area, highly uniform evaporated formamidinium lead triiodide thin films for solar cells. *ACS Energy Lett* 2017;2:2799–804. https://doi.org/10.1021/acsenergylett.7b00967.

631. Liu T, Lai H, Wan X, Zhang X, Liu Y, Chen Y. Cesium halides-assisted crystal growth of perovskite films for efficient planar heterojunction solar cells. *Chem Mater* 2018;30:5264–71. https://doi.org/10.1021/acs.chemmater.8b02002.

632. Yang L, Wang X, Mai X, Wang T, Wang C, Li X, et al. Constructing efficient mixed-ion perovskite solar cells based on TiO_2 nanorod array. *J Colloid Interface Sci* 2019;534:459–68. https://doi.org/10.1016/j.jcis.2018.09.045.

633. Park J Y, Park S J, Kwak M, Yang H K. Rapid visualization of latent fingerprints with Eu-doped $La_2Ti_2O_7$. *J Lumin* 2018; 201: 275–83. https://doi.org/10.1016/j.jlumin.2018.04.012.

634. Wang M, Li M, Yu A, Wu J, Mao C. Rare earth fluorescent nanomaterials for enhanced development of latent fingerprints. *ACS Appl Mater Interfaces* 2015;7:28110–5. https://doi.org/10.1021/acsami.5b09320.

635. He Y, Xu L, Zhu Y, Wei Q, Zhang M, Su B. Immunological multimetal deposition for rapid visualization of sweat fingerprints. *Angew Chemie* 2014;126:12817–20. https://doi.org/10.1002/ange.201404416.

636. Saif M, Alsayed N, Mbarek A, El-Kemary M, Abdel-Mottaleb MSA. Preparation and characterization of new photoluminescent nano-powder based on Eu^{3+}: $La_2Ti_2O_7$ and dispersed into silica matrix for latent fingerprint detection. *J Mol Struct* 2016;1125:763–71. https://doi.org/10.1016/j.molstruc.2016.07.075.

637. Saif M, Shebl M, Nabeel AI, Shokry R, Hafez H, Mbarek A, et al. Novel non-toxic and red luminescent sensor based on Eu^{3+}: $Y_2Ti_2O_7/SiO_2$ nano-powder for latent fingerprint detection. *Sensors Actuators B Chem* 2015;220:162–70. https://doi.org/10.1016/j.snb.2015.05.040.

638. Park SJ, Kim JY, Yim JH, Kim NY, Lee CH, Yang SJ, et al. The effective fingerprint detection application using $Gd_2Ti_2O_7$: Eu^{3+} nanophosphors. *J Alloys Compd* 2018;741:246–55. https://doi.org/10.1016/j.jallcom.2018.01.116.

639. Dhanalakshmi M, Nagabhushana H, Sharma SC, Basavaraj RB, Darshan GP, Kavyashree D. Bio-template assisted solvothermal synthesis of broom-like $BaTiO_3$: Nd^{3+} hierarchical architectures for display and forensic applications. *Mater Res Bull* 2018;102:235–47. https://doi.org/10.1016/j.materresbull.2018.02.003.

640. Girish KM, Prashantha SC, Nagabhushana H. Facile combustion based engineering of novel white light emitting Zn_2TiO_4: Dy^{3+} nanophosphors for display and forensic applications. *J Sci Adv Mater Devices* 2017;2:360–70. https://doi.org/10.1016/j.jsamd.2017.05.011.

Index

Note: **Bold** page numbers refer to tables and *italic* page numbers refer to figures.